NO REGRETS
How to foolproof your dating life

Copyright © 2015 by Issac Curry

Published by Godzchild Publications
a division of Godzchild, Inc.
22 Halleck St., Newark, NJ 07104
www.godzchildproductions.net

Printed in the United States of America 2015—First Edition
Cover Design by Ana Saunders of Es3lla Designs

All rights reserved. Except as permitted under the U.S. Copyright Act of 1976, this publication shall not be broadcast, rewritten, distributed, or transmitted, electronically or copied, in any form, or stored in a database or retrieval system, without prior written permission from the author.

Library of Congress Cataloging-in-Publications Data
No Regrets: How to foolproof your dating life/Issac Curry

ISBN   978-1-942705-06-2 (pbk.)

1. Curry, Issac. 2. Relationships 3. Dating 4. Faith 5. Singleness 6. Love 7. Men 8. Women 9. Christianity 10. Self-Help

Unless otherwise identified, all Scripture quotations in this publication are taken from the New Living Translation (NLT), New Revised Standard Version (NRSV), The Message Translation (MSG), The King James Version (KJV) and The New King James Version (NKJV).

All personal stories and/or references are true and names have been changed to protect the identity and privacy of each individual.

TABLE OF CONTENTS

Preface *1*
Instructions Not Included *5*

Section One: THE NEW NORMAL

1. The New Normal
Redefining The Meaning Of Dating *11*
What Is Dating? *13*
What Is The Purpose? *18*
Define The Relationship *22*
How To Decrypt The Mixed Messages *29*
Is He Really Into You? *32*
The Temptation To Settle *36*
You're Worth The Wait *43*
Five Questions To Ask Before Re-entering The Dating Scene *47*

2. The Temptation Of Impatience
Overcoming Our High Demand For Microwaved Relationships *51*
"Just Add Water" To Your Relationship *52*
Convenient Or Committed? *56*
Imaginary Enemies *59*
I Want That *64*

3. Dating Your Insecurities
Recognizing The Damage Your Insecurities Can Cause To Your Dating Life *69*
No Representatives, Please *71*
The Birth Of Our Insecurities *74*
Carry Your Own Burdens *76*
I Am Not My Insecurity *79*

4. The Unexpected Gift
Why You Should Embrace Your Season Of Singleness 87
Who Switched Your Price Tags? 89
Peter Principle In Romance 92
Defective Relationships 101 94
Choosing Me Before We 98
Closing Thought & Prayer 104

Section Two: RECYCLING BAD HABITS

5. You Don't "Need" To Date
Dating And The Crisis Of Codependency In Relationships 107
Recycling Bad Habits 109
Nourishing My Codependency 113
Casting Call 117
Jesus Deficiency 119

6. I Loved It When She Called Me "Daddy"
Dating And The Tangled Web Of Transference In Our Relationships 123
What Is Transference? 123
I'm The Daddy 126
Bad Habits Die Hard 128
The Elephant In Your Relationship 130
Healed People Heal People 138

7. Seven Reasons Relationships Go Bad
Avoiding The Behaviors That Sabotage Your Dating Relationships 141
Seven Things That Sabotaged Your Relationship 141
Closing Thought & Prayer 152

Section Three: GETTING PAST YOUR PAST

8. Opening Your Hurt Locker
When The Pain Of Your Past Disrupts The Potential Of Your Present 155
Close The Emotional Back Door   156
Remove Your Mask   159
Yard Sale   162
Friendly Fire   165

9. Conquering Your Comma
How Identify And Remove Yourself From Toxic Relationships   171
Misplaced Punctuations In Your Life   173
Teenage Mutant Ninja Relationships   175
It's Not You, It's Me. No, Really, It's Me   179
I Can't Be Who You Want Me To Be   182
You Can't Come With Me   188
Stop Ignoring The Alarms:
5 Signs That Your Relationship Is Toxic   195
Is This Relationship God's Will   201
Loyalty Is Not Love   204
Never Return To Egypt   206

10. Have The Funeral
Learning To Let Go And Move Beyond Your Past   223
Exit Stage Left   225
The Autopsy of Your Failed Relationship   227
Have The Funeral   231
Leave The Gravesite Behind   235
Five Signs Its Time To Move On   239

11. But What If My Wounds Never Heal?
Living Beyond Your Hurt And Becoming Whole   243
The 9th Step   *244*
Permission To Be Selfish   *247*
Time Won't Heal Your Wounds, Jesus Will   *250*
Closing Thought & Prayer   *256*

Section Four: CULTIVATING A NEW CULTURE

12. Press The Reset Button
Learning How To Start Over, Again   259
Redefining Reset   *260*
Facing Your Ground Zero   *262*
Back To The Basics   *266*
Closing Your Gaps   *268*

13. What's For Dinner?
Taking Sex Off The Table And Pursuing Real Intimacy   273
Is The Sex Bad?   *274*
The Morning After   *278*
Sex Rules   *280*
Playing With Doh   *284*
Sex As Worship   *288*
Closing Though & Prayer   *290*

Section Five: DATING WITHOUT REGRETS

14. I Know Who You Are But Who Am I?
Redirecting Our Search For The Right One And Becoming The Right One   293
Who Am I?   *293*
Destiny Or Bust   *302*
Becoming Mr. & Mrs. Right   *304*

15. No Regrets: 7 Foolproof Methods To Safeguard Your Dating Life
Living The Life That God Intends For You 307
Seven Sure-fire Methods To
Help You To Date Without Regrets 307

Closing Prayer And Final Words 317
About The Author 321
Notes 325

Dedication

Mom
Dora
Curry Family
Hulon Family
HSC
J. Jackson
E. Bresee
M. Ruff

PREFACE

I confess. The mere thought of writing and publishing this book scares the freaking daylights out of me. You are probably saying, "Not another dating book," right? I totally understand. But this is not another "dating" book. This is something a little more than a book about dating. This is about discovering your life's purpose and listening to God's voice; while along the way, learning how to effectively date without losing yourself and jeopardizing your relationship with God.

I remember teaching my very first singles conference about eight years ago that was hosted by my good friend Dr. E. Marcel. I was offered an opportunity to teach because all the other teachers for this particular conference had backed out - for whatever reason, they were not able to follow through with their commitments. It was about two weeks before the conference and I had to prepare to teach church people how to effectively live a single life. Although my feelings were a little hurt because I wasn't the original choice to be the speaker, I soon relieved myself of that emotion.

I don't take any teaching opportunity lightly, so I was petrified. The fear of being exposed paralyzed me. Of course I was single, but I was still a very young adult. I mean, I knew I didn't have it all together in my own life. My facade of self-sufficiency was about to bite me in my behind, and the people were about to learn not only did I not have all the answers, but I still had questions myself. "I am a fraud," I told myself. "What am I supposed to tell the people?" I think I went a couple of days without eating any substantive food because I was so weary. I know for certain that the day before the conference I may have lost a couple of pounds. Imagine that.

I think what plagued me most was that, not only was I still wrestling with what it meant to live a meaningful solo life, but the context from which I derived, I had no examples. I didn't grow up with a father so I didn't have the privilege to see mom and dad together to encapsulate for me what a healthy relationship looked like. No one in my family had ever been married, unless you count my uncle getting married at the halftime of a Chicago Bull's game, in front of the television, by a reverend they met on the street. Needless to say, that marriage didn't last - it started off all wrong.

The closest I got to observing relationships was the revolving door of toxic relationships that my mom had with men. While I was growing up, there was verbal, emotional, and physical abuse. In turn, in my early relationships, I repeated what I saw. For this reason, I figured I should be the last person to teach about singleness. I made all the mistakes. I knew all the games to play. I was manipulative. But all these things, God and others reminded me, are what contributed to my passion to listen and to truly learn how to live a more concentrated life for God.

Walking into that classroom to teach my very first singles conference, God comforted me and let me know that I didn't need to be perfect in order to share with the people what God wanted me to share with them. I allowed myself to be a vessel and God did the rest. Believe it or not, the title of that class was, 'Dating Without Regrets: How to Foolproof my Dating Life.' After I finished teaching, I remember countless people approaching me and asking, "Where can I buy your book?" Since then I've had people inquiring and encouraging me to write.

Preface

This book is dear to me; it has been five years in the making. I became impregnated with this word at the cusp of a failed engagement. I tried to ignore it. I tried to run from it. I tried to preoccupy myself with other things. I thought the deep felt need to compose and complete this project would go away. I tried to do everything but write this book. I even went back to graduate school and earned another degree. I even tried to get a PhD degree but God closed that door. I thought maybe I would just write an autobiography because it seemed easier, but God said no.

The truth is, I didn't know the first thing about how to write a book so fear crept in. That fear kept me from really focusing and giving birth to this assignment. I was afraid I had nothing relevant to say. I was afraid nobody would want to read anything I wrote. "Who am I, that people would listen?" I would ask myself. I was afraid of failure. But I can honestly say I was focusing on the wrong things because, ultimately, I can't control who reads this book. I switched the price tags between being successful and being faithful. Success is not defined by how many people read this book but whether or not I have been faithful in penning the words God urged me to write.

In *No Regrets*, I am unapologetically transparent about my life as I share my story with you. But this book is more than just an account of my life, it's about our lives, and about the lives that God desires for us to live. This book is about dating – but it's about more than just dating because dating is an end in and of itself. I offer this book to not only share more effective dating techniques, but also to encourage you to first discover God's will for your life. Through this discovery, you will become more fruitful and purposeful as you grow closer to God.

Why dating without regrets, you ask? Having traveled

throughout the US, various countries, and through a few countries on the continent of Africa, I've discovered the similar fate of failed and faulty marriages, abandoned promises, and broken relationships that plague more than just the community I pastor here in the US.

I have met with countless people bearing broken hearts, searching for answers, and looking for love. I have experienced and witnessed for myself so many people settling for below-par relationships because they secretly feel they are not worth better. Insecurities, fear, and impatience - robbing us all from a promising future with somebody that deserves us. I learned, it's not just a Western problem; it is a human problem.

I was raised in a predominantly African American church tradition, although, I now serve as a pastor in a multiethnic and intergenerational ministry. In all of my traveling and in my serving as pastor, I've learned that our desire to love and be loved and to find somebody that will accept us for who we are, transcends cultures. It transcends generational and ethnic barriers. However, our world today is inundated with conflicting views about how to be happy and content while single, and how to go about finding true love. How do we effectively date in a culture like today's without being consumed by regret? How do we date without sacrificing our fellowship with God in the process? That's the question this book intends to exhaust. Nonetheless, by definition, to "foolproof," means to make resistant to and prevent from failure. The purpose of this book, however, is to discover and help provide effective ways for single Christians to foolproof their daily lives as we choose to integrate our dating life into our spiritual world and not the other way around. There is no way possible you can effectively assimilate your spiritual life into your dating world because there will never be enough room for God. I've tried it and have experienced an epic fail each time. And to honor my first teaching opportunity, I ultimately decided to give this book the same name.

I'm very thankful for my supporters for encouraging me along the way. I am grateful for the community of which I am pastor, who have motivated me and given me the space to be able to focus on this work. It is my desire that you take the time to give this book a chance to speak to you, even as it has blessed me in the process.

INSTRUCTIONS NOT INCLUDED

Dating comes with no real instructions. What should happen is that before we enter into a dating relationship we should be handed a black, unmarked envelope that miraculously falls from the mystic iCloud and officially informs us that what we are about to do is at our own risk. I know it sounds morbid, but it's real talk. Dating serves its purpose but it indeed comes with the greatest of risks. Nobody is certain to come out without bruises, wounds, and/or nightmares that will perhaps follow you into perpetuity.

Sometimes I hate dating. Is it okay for me to say that? I mean, there is no cookie-cutter person and yet we are programmed like robots and taught, "We must find Mister and Missus. Perfect" (That was in my robotic voice). So we go through this production-like line, sort of like in a factory, moving from one person to another looking only for defects. We are looking for perfect people, when in fact there is no such thing as a "perfect" person. But because we spend so much time looking for that person we keep pressing through the production line moving from one relationship to another bypassing imperfect people that perhaps may have been perfect for us. There should be some type of manual that each person should be required to carry around with them, like a drivers license, that informs us on how to best operate the machinery, right? Maybe that's wishful thinking. Maybe nobody would consider dealing with me after reading my instruction manual; it will probably read in some unspoken foreign tribal language indicating that Issac is just too complicated to translate - OPERATE AT OWN RISK!

There are things I wish I knew before I started dating. I wish I had known that hearts were going to be broken, by me. That pain doesn't always go away, especially not with time. I wish I had known that she may hate me for as long as she lives. Or that you just can't win them all. I wish somebody had told me that you will have to make tough decisions that seem like neither is better and that you will appear to lose in all scenarios. Or that sometimes you will have to make a decision between two people, your past and your present and maybe, just maybe, your past would have been the best decision. I wish somebody had told me that when you give your heart away, sometimes you will never get it back, or that some people, no matter how good they look, are not deserving of your love. I wish someone had told me that living in regret is like imbibing poisonous hemlock juice—it paralyzes you from the inside first, then it corrupts everything else.

If I had known that turning around was not the best way to move forward I wouldn't have spent so much time going in circles. I've learned that dating is a dangerous sport when you don't know what you're doing. People get injured. Irreparable damage can be done. Lives can be lost. Your age or profession doesn't determine your qualification to date. I know people that are seasoned in age who are still making a mess of things. I know doctors and lawyers that can't seem to get it right. It's your ability to become a learner and a student of yourself. Understanding who you are, what shapes your character, what makes you tick, your identity in Jesus Christ, and God's calling upon your life, help to give you direction and guidance in the sphere of dating, connecting, and communicating with others. It helps to make sense of your purpose for dating in the first place. Some people live a lifetime, afraid of looking in the mirror to face themselves because they don't want to embrace their areas of their lives needing improvement, their frailties. When a person is willing to become a student of his or herself, their past, and God's will, they understand how to better navigate this world of dating and embracing others. You're not just blindly looking for someone to fill a void left by a parent or from somebody significant in your life. Your impetus for dating won't be a direct result of observing that everybody around you is getting married and they appear to be happy so you feel the pressure to follow suit.

If you could write a brief instruction manual about yourself that other people would be able to read, what would it say? Beyond the surface level things similar to what you would position in an online dating profile, can you articulate significant things about yourself that would help others to understand who you really are, where you're going, your honest areas that are needs for improvement, and how to get the best performance from you? I challenge you to include your own instructions.

Section 1
THE NEW NORMAL

"Attract what you expect, Reflect what you desire, Become what you respect, Mirror what you admire."
-Author Unknown

CHAPTER 1

THE NEW NORMAL

REDEFINING THE MEANING OF DATING

"How many times should you go on a date before you have sex?" Kenny lobbed the question into the crowd of men like a hand grenade into a crowd of unsuspecting civilians. The thing is, they were all standing around at a singles' ministry potluck dinner. I was sitting afar off acting as if I didn't hear the question being trumpeted, hoping that someone would speak up and at least set the record straight. "Should it be three dates or four? How long before it's okay to have sex?" he added. I didn't know if I was more shocked at his lack of discretion and awareness of his context or at his apparent forfeiture of standards to the dominant culture, but either way, this question pointed to something larger than him or me, or the single adult ministry gathering together for dinner. His inquiry spoke to the demise of courtship and the unregulated and boundless features of our new age dating game - the new normal. In our modern world of "dating," we have created a new culture. A dating culture that has influenced all generations from the Baby Boomers to Generation X and so forth. A culture that I am guilty of having heavily propagated in my past. A culture where texting has become the primary channel of communication in relationships. Where social media is the preferred choice of disagreeing and even breaking it off with your beau. A culture where you are limited to 140 characters or a tag in a photo to express your feelings or the proof of your fidelity to one another.

Where meeting at a designated place for organic conversation and face time is squelched and abandoned for a more distant and artificial connection behind some computer screen and a keyboard, and maybe a little digital facetime through a mobile device if you get lucky. This is a culture where living together without the commitment of marriage has become the new way of building a "modern family." It's no wonder that women today have put out with little to no expectations of a real relationship and most men worth their salt are apprehensive to make that "commitment." The rules to the dating game have become so convoluted and coded that even a skilled Cold War spy wouldn't be able to decrypt it. So say it with me, "Good riddance to the old dating culture, there is a new kid on the block."

Shelly stood there staring at me. Her eyes were watery but she gave her earnest attempt to act as if she was "okay." After exhausting her third dating relationship in a 14-month span and resorting to communicating via text message with the last guy she was considering to date simply because that was the most interaction she could get from him, she wailed, "He even had the audacity to send me a nude photo! I mean, we barely even talk outside of class, we don't talk much on the phone, I didn't even know we made it that far yet." She clamored with great frustration, "That's it. I'm finished. I'm through dating. As a matter of fact, the word 'date' should be expunged from Webster's dictionary. I just need to be by my damn self."

I sat there hurting for her because we had been friends for 20 years and I was there to help her pick up the pieces through each of her last failed relationships. She wasn't wrong for giving dating another try. She wasn't at fault for desiring to be desired or for wanting to build a family although she was beginning to race against her biological time clock - a woman's greatest nemesis. She was turning 34. She was successful, well-educated, beautiful, and independent. But she was still single. I understood her sobering struggle. "The paradigm has shifted. Dating is outdated. Hooking up in an estranged culture is here to stay," I thought to myself. But her situation and my own daily struggle with being single began to make me wonder, "What does 'dating' even mean to a believer, today?" What cost am I willing to pay? What distance am I willing to go? What am I willing to compromise? How do I navigate this cryptic world of dating without being consumed in regret because I failed to give it my best effort, because I held back due to the crippling fear of failure, because of the shame of my past, or because I failed to listen to the voice of God? How can I date without accumulating any real regrets?

WHAT IS DATING?

I was invited to join a conference for pastors and church leaders, which was hosted by a very popular and respected elder in a nearby city. It was an invite-only engagement so I assumed this would be a very special moment in the life of my ministry. At this conference there were leaders, ministers, and elders from around the country. Pastors who were considered more experienced and who were excelling in ministry and leaders who were seeking wisdom about ways to grow and excel in ministry; all gathered together under one roof. As the conference came to its culmination, there was a discussion panel designed to provide answers to very difficult questions for all pastors who were seeking direction. If you had questions, they provided answers.

There I was, sitting among a consortium of pastors with a panel of decorated leaders sitting before me tackling tough issues and questions, left and right. Then it happened, somebody removed the pin from the grenade and politely sat it on the center table, daring another to reach for it, "If a pastor is single, how can he or she date?" I gulped. I mean, the fact that the question was even asked made me think that I had clearly been riding the wrong train for 15 years. "You mean to tell me, I'm not supposed to be dating because I'm a man of the cloth?" I thought to myself. I was clearly one of very few pastors in the room still walking around with an "S" on my chest, like I was a Single superhero. I mean when I walked, I could sometimes faintly hear theme music in the background something similar to Superman - the knockoff version. There was a noticeable silence across the room. Then eventually one panelist grabbed the microphone and said, "Well, for starters, we need to define 'dating.'" "No duh, Sherlock," I whispered to myself, which I occasionally do. I adjusted myself in my seat because I knew this was about to get very interesting. They went back and forth for about fifteen minutes on what dating was or what it was not.

"Dating is sort of like a prolonged interview," I heard one say. "Dating does not mean to sleep around," another added. Few gave their interpretations of what it meant and how it worked for them in their context, while others took a different approach because they felt the question didn't apply to them. Nonetheless, the subject was inadequately dealt with. From the audience's perspective, it appeared that each pastor who had the courage to speak on the subject owned a different perspective. Needless to say, I left that consortium with many things pervading my thoughts, one of which was about dating and being a pastor who is single. If the pastors who lead various congregations around the country had extreme difficulty coming to a

Section One: THE NEW NORMAL

unified position of dating and what it means, what then does that say about the parishioners and people they lead in the church that are still living single? Somebody is in trouble. This is not to bash the pastors who I represent. This is just to acknowledge that obvious variations of this term exist and what "dating" even means in the twenty-first century is complicated to discern. I was very compelled, so much so, that I returned home and installed a brand-new Bible study series called, "Grey Matter: When things in the Bible are not in black and white."

I was challenged in this series to open the floor for biblical and practical discussion about subjects and matters that, in the religious community, don't always get discussed due to the fear of not having an adequate answer. Subjects about love, dating, sex, abortion, sexual orientation, etc... all were up for grabs. In all my years of preaching, this series stretched me the most. It forced me to look at myself, my life, and my truest convictions. I had many questions come in from many different angles, and because I asked for the questions in advance, I had the opportunity to review them in order to determine my course of response. I remember unfolding pieces of paper to read the questions I was up against; one read, "Is it okay to have friends with benefits?" That question did not surprise me. It was sort of expected. Another read, "If you're in the throws of a divorce or in the process of being separated, is it okay to still date?" I figured the writer of that question was specifically referring to somebody he or she knew personally and wanted me to be the one to address it on their behalf - anonymously that is. But there was one particular note, entitled, "Relationship advice." It read, "I'm dating this new guy and I really like him, but he told me that if I wasn't willing to have sex with him, he wasn't going to waste his time getting to know me. So I am worried that if I tell him, no, he won't want to date me anymore." Now, this was totally unexpected. This caught my attention and wrenched my emotions at the same time. No name. No source. Just a real life dilemma.

This piece of paper spoke to the tragedy of the now. So often, we find ourselves in relationships where we are manipulated and coerced into giving up who we are, compromising what we believe, or overcompensating for something we don't have or feel we must obtain just to vie for the approval and appreciation of the other person. "How do I go about addressing this question?" I pondered. In past years, women would fret over whether or not giving up a kiss on the first date sent the wrong message. Now women fret over whether having sex on the first date will keep a man interested enough to stick around. Dating is such a lost art. We must, for the sake of restoring sanity, raise the standards to this fiasco.

So Sunday arrived and we began our first installment of, "Dating, Sex, and Relationships." We had a full house of single adults with a few married couples sprinkled in the audience - everybody ranging from 30 to 75 years of age. I remember approaching the podium and, first, with a melodramatic hue, I asked the question, "What is dating? How do we define, dating?" The entire room was at a complete stand still. Nobody could think of a substantial way to answer what most people consider to be a very simple term. Go figure. So imagine, entering into a "dating" relationship with another person without there being some semblance of clarity to the questions 'what are we doing here?' and 'where are we going'? and yet expecting to achieve a common goal that perhaps the other person never even considered to be an option.

A very critical factor that contributes to the reason most dating relationships fail is that the two people aren't on the same page. So one or both parties are left holding bags filled with unexpressed and unmet expectations. Assumption is like poison, just a little bit can taint any relationship, I don't care how promising it may be. I know that story all too well. So, for the sake of clarity, I began defining what it means to date as "building a friendship and relationship with another person with the hopes and intentions of determining their suitability to be your life partner." The ultimate purpose for dating should be to retrieve the necessary "data" from a person in order to make an informed decision about whether you should invest more of your time, your energy, and your emotions into that person or a relationship with that person. If your destination is something other than this, you're playing the wrong sport, bubba. But no matter the nuance of your definition or understanding, this is a conversation that you want to have at the forefront, before any time has been spent, decisions made, or boundaries have been crossed. Most of us men, or at least I know, as I reflect on my personal past, have mastered the art of maneuvering "the conversation" and most women procrastinate initiating "the talk" until after the damage has been done and the relationship is on its downward spiral. This is simply because you assumed he felt the same way about you. As I began wrapping up the bible study, Sarah, 54, cautiously walked toward me and with a hybrid of embarrassment and confusion written across her face, she whispered, "Did you get my note? I was too afraid to write my name on it in case it fell in the wrong

> **NR**
> Dating: to build a friendship and relationship with another person with the hopes and intentions of determining their suitability to be your life partner.

hands." I looked her square in the eyes and I told her. "Simply put, he is toxic for you." I told Sarah, and now I am telling you, the purpose of dating anybody is so that you can get to know him or her better with the intentions of determining their suitability to become your life partner period. If at the beginning you are given stipulations to have sex or any other thing you sincerely feel compromises who you know yourself to be, or if you have to jump through too many hoops just to find clarity about the direction of your dating relationship, then let the search continue they are not for you.

WHAT IS THE PURPOSE?

I remember growing up in Anaheim, California, as a child and having to move abruptly to Memphis, Tennessee. My mother wrote a farewell letter to my stepfather and while he was at work she took my sister, my younger brother, and me and we caught the greyhound bus and rode halfway across the country. It was very confusing at the time because everything was seemingly going well but we were too young to completely understand what was going on. We were forced to abruptly leave behind all we knew. However, my sister and I learned when we got older that there was really nothing wrong, my mother just needed a change of scenery from a stable and functional 10-year relationship. At a young age, I was taught the power of flight in relationships. That is, running away, commitment-phobia, insatiability, boredom, and a recycled dysfunction of finding or exacerbating a problem when little or no problem really exists. This is the hand I was dealt. I didn't realize how much these pivotal childhood moments were crafting my understanding about love and relationships until I found myself mirroring them in my own life.

Section One: THE NEW NORMAL

I dated until the newness of the relationship dwindled and I got bored. I would then find something I didn't like and I created escape routes to exit the relationship like an interstate highway. I know, it was bad. I thank God for His forgiveness for the messes that I have made. I later realized that I never really saw a marriage certificate, a wedding ring, photos of a happy couple, or what people call marital bliss. I was only exposed to the recycle bin of dysfunctional relationships and a revolving door of men that proliferated them. I had snapshots of physical, emotional, and psychological abuse of men toward my mother and sometimes she initiated them. When I started going to church and building friendship circles with married couples, I began to see the normalcy of infidelity and broken promises there, too. And with my single friends, I observed how they could sometimes imitate marriage better than my married friends just without the complications of being "married." I used to tell myself this is as good as it gets. "Why date? Why get married just to get a divorce?" I bull-horned in the privacy of close friends. "What is the purpose?" I assumed this was the end goal of relationships, and this is as close as it would come to being truly happy. At one point, I sincerely believed there was no hope or purpose for marriage anymore. For me, marriage had lost its luster; it was a thing of the past, something that older generations were able to bask in, but not today. The new normal, and one of the paramount difficulties today, is that single people are living like they are married and married people are acting like they are single. But thank God for the transformational power of the Holy Spirit who was able to change my entire way of thinking about love, relationships, marriage, and the will of God before it was too late.

I was on the verge of sabotaging my life with my foolish

thinking, but God had mercy on me. I found myself later on in ministry pondering the question, "What is the purpose?" And I remember God reintroducing me to a familiar text in Genesis 2 that helped me to understand that relationships, love, and marriage were not necessarily created for you and I, but for God, first. As I began to vigorously seek an answer to the question, "What is the purpose of marriage?" God was helping me through pivotal circumstances in my life to rephrase the question to "What is my purpose in marriage?" God began helping me to clearly understand that I was replicating the wrong formula to solve the equations in my life and in my failed relationships. I understood simple math to be $1/2 + 1/2 = 1$. That is half plus half equals one whole. I was not whole and yet I was seeking to find another person to complete me instead of compliment me. "No problem can be solved from the same level of consciousness that created it," says Albert Einstein. I spent my life copying incorrect behaviors from others thinking I could somehow change the results, but to my discovery, that wasn't working. The goal, for me, was to discover my purpose and to find fulfillment in God's will, first; otherwise, I would always have an insatiable appetite and continue to operate in dysfunction and fear.

In Genesis 2:8-25, Adam and Eve struck a nerve and forced me to take a fresh perspective of both their lives. Adam was created by God to not merely exist but to be fruitful and to glorify God with his life; to first live with purpose. The text says, *"God placed man in the garden of Eden to cultivate it and guard it."* *God later says, "It is not good for man to live alone. I will make a suitable companion to help him."* But the unique thing with the text is that although God acknowledged that it was not good for man to live alone and that he would make somebody that was suitable for Adam, there was still a gap between the statement that

God makes and when God actually brings that person into Adam's life. What's notable is that the "in between time," after verse 18 and before verse 21, Adam is found living out his purpose and fulfilling God's call upon his life. In this space, the Bible says: *"So God took some soil from the ground and formed all the animals and all the birds. Then he brought them to the man to see what he would name them; and that is how they all got their names. So the man named all the birds and all the animals; but not one of them was a suitable companion to help him."*

We don't really know the time lapse of this gap, we can only assume. But one thing is for sure, Adam was concentrating on what God was doing right in front of him. I imagine it had to take quite some time to create names for every bird and every animal. This is what God asked of Adam, and Adam was faithful in doing what God had asked him to do. Then, the Bible teaches that God caused Adam to fall into a deep sleep and there he made a suitable companion for Adam. The "in between time" where God recognized that Adam was alone and when God actually brought a suitable companion into Adam's life, was not filled with impatience, indifference, or complacency, but it was filled with purpose, identity, intimacy, and fulfillment. The "in between time" wasn't marked by a number, a date, or a year, but it was signified by faithfulness and obedience. Adam didn't become preoccupied with marriage or having a life partner but his main goal was responding to the voice of God, answering God's call, and achieving closeness with God. Becoming intimately involved with God was Adam's pursuit.

Looking at Adam and Eve with fresh new eyes, God revealed to me that marriage should not be the final destination for us to build our lives upon one day achieving, but a sincere intimacy with God should be our truest goal. This shifted my paradigm on relationships.

Adam and Eve teach us that marriage is about individual wholeness and about two people coming together and celebrating God together as One, and as One, your goal is intimacy not only with one another, but with God. PUBLIC SERVICE ANNOUNCEMENT: if you cannot find peace, happiness, fulfillment, and identity in God as an individual you will not find it with a marriage certificate. Adam was found doing what God called him to do, fulfilling his purpose. In due time, Eve came alongside him to accentuate what Adam was already doing with his life, and Eve was going in the same direction, serving a similar purpose with her life. This is not meant to sound cliché, this is truth.

> If you cannot achieve intimacy and contentment in God's will as an individual you cannot expect to manufacture it with somebody else.

Adam's single goal as an individual before marriage was intimacy and satisfaction in God's will for his life. If you cannot achieve intimacy and contentment in God's will as an individual you cannot expect to manufacture it with somebody else. If discovering your purpose and living in God's will is not priority, when you find a mate you can't expect for it to become priority. The most common yet imbalanced thing one single person can do is to enter into a relationship and focus solely on the other person's relationship and intimacy with God and not have an identity of their own. This is how insecurities are birthed (We will talk about that in chapter 2).

DEFINE THE RELATIONSHIP

I remember back in my early college days, I grew to be a typical frat boy. I became enslaved to my ego and I was a full-time employee of regret. I used my recently failed relationships during my sophomore year as a crutch to lean on and a tool to control the direction of any future relationship I engaged in. My commitment-phobia was sported like a newly purchased sweatshirt around campus, and it was stylish. I medicated myself with series of brief, short-lived, and passionate relationships that ended in haste. I had so much to learn about God's mercy and grace. I mastered the art of teetering the threshold of committed relationships but yet somehow remaining "just friends." I stood on the edge of broken hearts and peered the campus at the sea of unmet expectations that I was single-handedly responsible for. I learned to carefully orchestrate manipulation by pretending we were never more than friends when in fact, we were. I learned the art of acknowledging at the onset of any would-be-relationship that, "I was not ready for" or "I was afraid to get my heart broken, again." I carefully set the stage so that we could define the relationship as nothing more than friends. However, my actions were never really congruent with my words. This way, I could easily point the finger when emotions flared and the expectation of something more was mentioned. The subject of a committed relationship became taboo. It was manipulation at its best.

Have you ever been in a situation where somebody's actions and their words did not intersect? What did you do? Have you ever had a relationship that if I asked, you wouldn't be able to clearly and confidently define "what" you are and "who" you are to one another? How did you respond? Can you define the relationship you're in right now? I mean, he says he isn't ready for a relationship but somehow you two spend time doing relationship-type things together. He still somehow manages to send flirting gestures toward you when you see one another, and the time you spend together is without question, prime real estate time. You've asked yourself over and over, "Where is this going?" You've come to the conclusion that you want more than what you're getting, so you do the unthinkable and decide to go for the jugular and express your feelings to him. Now that you've told him how you feel it seems like he slowly drifts away or appears emotionally removed from the "friendship" - it's like everything began to go downhill when you decided to do what normal people do, express your feelings.

Have you ever felt like you sort of got punished for wanting or expecting something more to come from your "just friends" relationship? Have you ever been in a "friendship" but it felt more like you were exclusive? What's the deal with the "Friend Zone" anyway? I'll tell you. In most cases, it's a tool of manipulation. It's a barrier of comfort and an asylum in which the initiator retreats and is exempt from obligation, accountability, and/or emotional responsibility. When you or the other person's actions communicate that you are more than friends but your words say we are, "just friends," you are either lying to yourself or you are intentionally deceiving the other person.

Have you ever experienced this before? Were you dating somebody, or at least it "appeared" that you were dating but somehow you could not evade the category of just a "friend?" How did you deal? What did you do? There are some common relation — slips, simple mistakes that hinder us from being able to move to that next place in our relationship.

1. Many times the mistake comes with our unwillingness to define the relationship or to get clarity of the direction of the relationship from the very beginning. With our unwillingness to define the relationship, we also fail to speak to the incongruences and call them out when we

Section One: THE NEW NORMAL

see them. That is, if you are just friends, it shouldn't matter if you decide to go out on a date with another person. If you're only friends, he or she shouldn't get upset if you fail to answer or return a phone call.

2. Other times the mistake comes when we fail to develop boundaries and we choose to become "physical" and among other things, we treat our friendship like it's a marriage without the commitment. Hypothetically speaking, if being single I am able to have sex with you and to live and operate with you as if we are married, what then is the purpose of getting married? The reason so many relationships fail to move beyond boyfriend and girlfriend is because one or both persons have

Many times the mistake comes with our unwillingness to define the relationship or to get clarity of the direction of the relationship from the very beginning.

everything they need without the commitment. Because of this, a man, or woman for that matter, loses a sense of urgency of getting and wanting to know you more intimately. Without developing boundaries, it becomes practically impossible to separate the sincere from the fake.

3. Another essential mistake comes when you give him the impression that you "need" him through your actions. When you're only friends, but even if you're more than friends, you should never make the mistake of revolving your world around him or her. It's not always a good idea to allow a person too much access to you. Sometimes it's not wise to become too available for him. Any person can sense when they have become your all in all, and it usually results in the other person's pulling back or taking advantage of your open heart.

DOES HE WANT YOU OR WHAT YOU CAN DO FOR HIM?

I remember receiving a letter from a member and friend who found herself on the short end of an undefined relationship. "I feel I am just a fool when it comes to dating and relationships with men," Carlisha confessed. "Paul and I have been friends for almost three years now. There was a vibe between us that was undeniable and everybody saw it. Paul's voice was the last voice I would hear at night and my voice was the first he would hear in the morning. We continued to say we were 'just friends,'" Carlisha explained, "but it seemed like we were much more than that." "We never had sex but we did cross certain boundaries, if you know what I'm trying to say? I feel awful about it because He knew I didn't want to have sex but we always seemed to find ourselves in those situations where we were pressing the envelope."

In this letter, Carlisha continued being vulnerable with me:

"It's embarrassing to mention but Paul is still married although he has been separated from his [wife] for 8 1/2 years, now. He never really sought to follow through with making the divorce final. In his mind, he stressed that he considered him and his [wife] to be divorced and since he had no desire to remarry, he saw no real significance in seeking a divorce. He said it would cost more than he could afford. He told me over and over he would never date anybody ever again so I stopped pressing him about us becoming something more than just friends. I accepted that we would be nothing more than friends but we continued to spend time together as if we were actually in a dating relationship, and he continued to pursue me physically, but to no avail. He was actually spending a late night over my house watching movies just one week ago when I finally told him 'we are not going to have sex.' I expressed how I really felt about him. I also asked him if he could use some financial assistance to help get through his divorce. He declined my offer but he seemed okay about me sharing how I felt about him. Eventually he left. I haven't heard from him in a few days which is not normal, so I thought I would text him and ask him to come and sit with me during Bible study as he often would do. His response startled me, and really, if I can be honest, it broke my heart. He told me that he couldn't because he started dating Patricia Pasmos, another church member.

Section One: THE NEW NORMAL

I could tell Paul was starting to pull back but I didn't think he was capable of being so insensitive. I am so confused. Was it something I did? Was I not enough? Where did I go wrong? I took the risk of expressing my feelings to him and this is what I get in return? I mean, how can he be so adamant about not wanting to date and not getting a divorce and now he's openly dating someone else and he's going through the process of completing his divorce?"

I told Carlisha, "You are not crazy. You are certainly not a fool." Sometimes we give people the benefit of the doubt when we know they don't deserve it. Sometimes we can want something to work so badly that we don't pay attention to the red flags. Sometimes you have to stop and ask yourself, "Does he really want me or does he want what I can do for him?" I shared with Carlisha the inconvenient truth that Paul obviously only wanted what she could do for him.

"You could feel his voids. You could provide for him an escape. You could make him feel good when he was in need. You were available to talk to him when he needed somebody to listen. You were comfortable but you were not who he wanted ultimately to be with." It's heartbreaking, but it's the risk we take when we ignore the signs and warnings that someone is not completely invested in the relationship. Sometimes in relationships we entertain and hang onto people who may not be

Sometimes we can want something to work so badly that we don't pay attention to the red flags.

marriage material. We keep them around just in case we don't find somebody better. It's your responsibility to make sure he or she not just keeping you around in case other pursuits don't work out or until somebody better comes along. In other words, don't allow yourself to be someone else's insurance policy. If you're not worth their first place then you're in the wrong place.

When you are trying to define your relationship be aware of:
- The one who tries to use his or her former relationship as an excuse not to engage you or your current relationship.
- The one who avoids introducing you to his or her family, key friends, and coworkers.
- The one who avoids making plans with you, or who often cancels at the most inopportune time and for ridiculous reasons.
- The one who only wants to spend time with you in private places, closed rooms, and only in the evening.

If you are "just friends" here are a few pointers:

1. Watch your conversation.
Stop with the innuendos. You can't get upset if he pursues you in sexual ways, if your conversation gives him that perception. Don't give room to it.

2. Manage your time.
Stop being so accessible. Stop giving up all of your time. Just because the "friendship" begins doesn't mean your life and goals has to end or even take a backseat.

3. Limit your touch.
Stop thinking the only way to show and express your adoration is if you become physical. Stop allowing your relationships to be sabotaged by a distorted view of intimacy.

4. Leave room for God.
Don't allow yourself to become so consumed in the newness of the relationship that you so easily close God out. It happens. Allow God to direct you, give you guidance, and to reveal to you whether you are in the right place or not.

HOW TO DECRYPT THE MIXED MESSAGES

Is he serious about me? Does she just want a fling? How long is he trying to date? Like cracking Soviet ciphers and interpreting morse code, you have been trying to make sense of his encrypted messages. You've probably realized by now that you have been wasting your valuable time playing a game of charades hoping that someday you would accurately interpret the mixed signals she has been sending you. Believe me, you deserve to know. You deserve clarity not confusion.

I've read many articles, blogs, and have done a lot of research on this subject. Some of the things out there right now on this subject are downright amusing. No offense. I read from a writer that although the "game" of mixed signals is more insecure and unsteady than a regular courtship, you shouldn't rush in looking for answers but its better to "take your time" and "enjoy the ride." Sorry, but not everybody wants to play games, especially the people I encounter on a daily basis. If you know somebody is sending mixed messages, I don't think you need to be a part of that game; furthermore, you need to get out of that vehicle because that is a ride you don't want to be taken on. I viewed on one site where a person offered advice on how to read mixed messages. It says something like, play along, challenge the game every now and then, and remember mixed signals are mostly sexual and that you need to create memories to excite their sexual side. In other words, it would behoove you to contribute to the foolishness, as if that will help with the confusion.

The Bible teaches us that when a man is confused it's because "their loyalty is divided between God and the world, therefore they are unstable in everything they do." (James 1:8) Mixed messages are not only a sign that there is likely a conflict between his emotions and his logic, but there is also a conflict with his spirit and his fellowship with God. Simply put, when a person is sending mixed messages it's likely because he or she has a deeper disconnect with their power source, God.

I likened it to my recent job as a leasing consultant of some local apartments. I had a responsibility to remain in constant contact with the agents in the field who were responsible for handling maintenance issues and maintaining the upkeep of the property. One evening I failed to connect all the walkie-talkies to their chargers to be charged overnight. Needless to say, the next day when I proceeded to use them to communicate a maintenance issue to my coworkers, they could not fully comprehend what I was saying or the message I was conveying to them. It was confusing to say the least. It wasn't their fault. I failed to connect the radios to their power sources so they could be fully charged and responsive. I believe its true, when a guy sends mixed messages it's because he has mixed feelings. However, when mixed feelings are at the core, it usually points to the much more significant reality that he hasn't been connected to the Source, or he isn't fully listening or following the direction of God. For this reason he is going to be confusing.

> **NR**
> I believe its true, when a guy sends mixed messages it's because he has mixed feelings.

I have found myself in many scenarios throughout my life where I have been the one sending mixed messages to a woman. Not because I truly set out to hurt any person or to be a jerk, but my mixed messages pointed to something more significant. During these encounters, if I am honest, I wasn't on the same frequency with God or seeking God's will. Perhaps it was my selfish ambition or a result of my trying to operate from my own will. I believe that when a person sends mixed signals, it's indicative of their personal life and emotional stability (or the lack thereof). It's a red flag. RUN! Or you can stick around and hopefully make him commit. Tell me how that works for you. Many times, the person sending the signals can be just as confused as you are, but believe me, his confusion is helping you to dodge a bullet. Don't allow yourself to be engulfed in confusion. While a man, or woman for that matter, is permitted to be unsure about whether he wants to be committed to you or about your future together, he shouldn't have the privilege to play house with you until he can decide. Whatever the reason, you deserve somebody who is ready. When a man is being confusing, you don't have to stay. When a woman is being confusing, you don't have to stay. You don't have to engage in the drama.

Here are a few steps to help you navigate the mixed messages you have been receiving.

STEP 1
Are you contributing to the confusion?
Be careful that you have not allowed him room to experience what he would get in a committed relationship without actually requiring that he commit to you. It's very hard to play committed with someone and then later ask him does he want to commit to you. Why would he commit to you if he has already had the luxury of having you to commit to him without having to reciprocate? Better yet, it could be that he has already sent you the clear message that he doesn't want to be with you but you refuse to take the message at face value. Perhaps he doesn't want you, only what you can do for him.

STEP 2
Accept the fact that he is unavailable.
Emotionally, Physically, and even Spiritually. Stop wasting valuable time, energy, and prayers on a person who doesn't want to be there. Stop trying to make him into something he doesn't want to be. Come out of denial. Accept that this just isn't where God is prospering you.

STEP 3
Give them space to be confusing by him or herself.
If he is constantly leading you on and not being clear with you about his hopes and intentions for your relationship, leave him be. It's not easy. I get it. However, in the long run you will be grateful that you allowed him to remain confused by himself or whomever he wants to be with. Back off. Give him space to be confusing without you in the picture. Give yourself credit, you deserve to have clarity.

IS HE REALLY INTO YOU?

I know this may sound ridiculous and very adolescent, but the rudiments of my understanding about love and relationships also came from some of my favorite childhood cartoons. One of my all-time favorites is Warner Bros. Looney Tunes, specifically, the Chronicles of Pepé Le Pew. For starters, Pepé, is a French skunk who is typically found strolling the streets of Paris in the spring time when everyone's thoughts are about love. There are four significant things you should know about Pepé, if you don't already. First, he doesn't realize he is a skunk and he has a bad smell. Second, he is blissfully convinced that a black cat is a skunk so he spends all his time and energy trying to make himself believe that she is something she is not. Third, he believes the cat is really into him therefore he avoids reading the signs and having to accept that she doesn't really like him. Fourth, he wants love so badly that he is willing to do practically anything to possess it and to get this cat to reciprocate the same feelings.

Pepé and other characters have placed their tendrils deep within my psyche. They taught me that:

1. When I see a woman that I like I need to chase after her until she gives in.
2. If I can just find one thing that we have in common that is good enough for a commitment.
3. Don't take no for an answer. No actually means yes.
4. Give everything that I have, deplete myself, even if she fails to reciprocate the same emotions and actions.

Ridiculous, right? It's true, though. This is how so many of us operate. The irrational romantic narrative portrayed in these cartoons are still the same scripts replayed throughout our culture, and because of it, everybody has their own warped view of romance and the ingredients needed for a healthy relationship. In this cartoon series there is an undeniable chase. There is an omnipotent desire to find love by any means necessary, even if it means ignoring the signs that the other person really isn't right for you. There is one episode where Pepé finally realizes that Penelope is not really like him - he discovers she's a cat when her stripes wash off. But because Pepé is so consumed with the idea of finding his "l'amour," he remains willfully blind and is undeterred so he proceeds to cover his white stripes with black paint, taking on the appearance of a cat before resuming his pursuit of love. Even if it means not being true to himself, he is willing to enter into a commitment with Penelope.

But now that I am older and after having tried to intently follow each of Pepé's rules for happiness and having failed terribly, I can look at those same cartoons and like hieroglyphics I can decipher what we really should gain from them.

1. Willful blindness and the overbearing desire to find love will cause you to ignore the signs that a person really isn't into you or a good fit for you.

Willful blindness and the overbearing desire to find love will cause you to ignore the signs that a person really isn't into you or a good fit for you.

2. Just because a person compliments you doesn't mean he or she values you.

3. Just because you appear to have some things in common on the surface level doesn't mean you share the same values. He was a skunk and he saw a woman that appeared to be a skunk and he knew she was the one for him.

I've been asked by so many, "Why isn't he responding?" "What do I do?" "I don't know if she really likes me," and the list goes on. Well, for starters, if you're doing all of the chasing and pursuing and there is no reciprocity you might want to consider another option because you're playing the skunk and he/she is playing the cat. Secondly, if you have to continue to ask yourself over and over whether and not the other person really likes you, that might be an alarm that you need to consider paying attention to. Thirdly, if you find yourself denying who you really are and not being true to yourself in order to receive acceptance and affirmation from the other person, it might be that he or she is not really into you, neither are they a good fit for you. The best thing you can do is to accept that a person is not worth the investment of you depleting yourself in order for them to be happy or content with you.

Let me put it another way.

1. Stop compromising your values and your goals just to appease him, satisfy his insecurities, or to get his attention.

2. Stop trying to fix everything. Stop working overtime trying to make things work. If she isn't putting forth a real effort to make the relationship work, then you shouldn't be busting your chops trying to overcompensate for her indifference.

Stop initiating the conversation about your future together and pleading for a commitment. If he really values you then he would not want to lose you.

3. Stop initiating the conversation about your future together and pleading for a commitment. If he really values you then he would not want to lose you. Playing games and sending mixed messages isn't the road to commitment but it's the path you take when you're in a holding pattern. Stop going in circles.

So many times I counsel and meet people who have spent their time, energy, and resources on people and relationships only to realize they have wasted it because the other person was never worth it. If the other person is not willing to be clear and transparent with you, don't waste your time playing charades hoping you will somehow break the code to her encrypted behavior. Remember, if you have to fight and struggle just to get him to commit to you, you're going to have to keep fighting and struggling to get him to stay. That's not the life you want to live. Let him or her choose you.

THE TEMPTATION TO SETTLE

"We are half hearted creatures, fooling about with drink and sex and ambition when infinite joy has been offered to us, like an ignorant child who wants to go on making mud pies in a slum because he cannot imagine what is meant by the offer of a holiday at the sea. We are far too easily pleased." When I think of these gripping words written by renowned author C.S. Lewis, I can't help but to look at my life and to think that for so long I've allowed myself to settle for mediocrity. Settling for less than you're worth is not only an action that needs to be corrected, it is a mindset and a cycle that needs to be broken.

My very close friend from elementary school sent me a text message one evening not long ago, it read, "I am pregnant!" That's it; nothing else came with the announcement. It was like she threw a Hail Mary football pass into the air and hoped for a touchdown. I replied, "Are you all getting married?" She said, "Eventually." Then there was this dead, awkward silence. She and I both knew that her "eventually" really meant, "I don't really know. We haven't talked about marriage." I didn't respond any further because I needed a little time to digest what I believed was taking place, *"Like a child who wants to go on making mud pies in a slum because she cannot imagine what is meant by the offer of a holiday at the sea. We are far too easily pleased."* A few days passed, she called me and wondered why I never replied back. She was bothered and could tell that I wasn't ecstatic although I was indeed happy for her. I was conflicted; I was happy but I was sad at the same time.

Because her previous relationships had failed, she had seemingly given up on marriage and given in to the temptation to settle for less than her value and less than what she desired most, a husband and a family. She reached her lesser goal instead, pregnancy. She said to me, "You know I want a baby more than I want a relationship, so if the relationship doesn't work, 'so long, see you later'," she referred to her relationship with her new boyfriend and soon-to-be baby's father. Truth is, she wanted a relationship more than anything but she was willing to settle because she had grown accustomed to disappointment and failure with romance.

I recently heard a message preached by Steven Furtick called the "Expectation Gap." He talked about how in our lives our expectations don't always translate into what we experience and for this reason frustration and disappointment fill in those gaps where our expectations and experiences fail to meet. If you're breathing, chances are you've been let down before, right? Maybe you got married with the expectation of living happily ever after, but after betrayal and your pockets filled with unmet expectations, you realized you got something different from what you expected in your marriage. Maybe you decided to give love and romance another chance and after giving your heart away you ended up on the short end of the stick of another failed relationship. Perhaps, you just seem to attract all the wrong men as if the fragrance you wear only gets the attention of men who don't want a commitment. Don't you get tired of being disappointed, dissatisfied, and left with unmet expectations?

Through my own life experiences, I have often felt that the best way to close my expectation gap was to lower my expectations so that they could meet my experiences. Does that make sense? At one time I thought it did. My temptation has been to simply "settle for what I can get" because sometimes it just seems that my personal preferences and my expectations have been too lofty. But I have met way too many people and have heard too many stories from friends, family, and colleagues about how they wished they never got married because they knew that person wasn't the best for them. Now they stand on the other side of a divorce. I've witnessed many people, who just like me, have wasted too much time babysitting relationships that were not up to par and trying to hold on to people that were not worthy of their time simply because they had given in to the temptation to settle for less.

Believe it or not, our temptation to settle for subpar relationships is just as real as our temptation to settle for deficient shopping carts when we are shopping for groceries. Humor me for a moment. There are four types of grocery carts that are very similar to four types of dating relationships you will encounter.

1. You have the Pusher

This is the cart that requires more of you than it should because neither of the wheels on the cart will move on their own. Therefore, from the very beginning, you can see it's an imbalanced cart-to-shopper relationship. These are the relationships that drain you because they require you to do all the work. These are the relationships that are imbalanced and you find yourself carrying the weight of the relationship on your shoulders alone. Don't settle for the Pusher.

2. Then you have the Sidewinder

This is the cart that only wants you to veer one way, with no compromising. So no matter where you are in the store or what you are trying to accomplish, the sidewinder cart has the tendency to pull you in one direction. If you're trying to go right and the cart has a predisposition to go left, you have to spend your time manipulating it so that you can get simple goals accomplished. These are the relationships you find yourself in where it is never about you and always about them. In the Sidewinder relationship you almost never get your way; there is never any compromising. The relationship revolves around the other person. Don't settle for the Sidewinder.

3. Next is the Noise-maker

This is the cart that you typically end up with after you have encountered the first two carts and returned them both. You notice with the noise-maker that it isn't perfect so you try to justify to yourself that the wobbly noise isn't that loud. That is until you hit your stride and when you do, you notice that everybody in the store has this "embarrassed for you" stare because you have the one cart that everybody else knew to avoid because it makes too much noise. These are the relationships and people you try so hard to justify your reason for staying. In the noise-maker relationship, you spend so much time making excuses for the behavior of the other person. They are belligerent when they get mad and you're often embarrassed and not happy in these relationships. Don't settle for the Noise-maker.

4. And finally you have the Faker

This is the cart that when you begin pushing, it seems perfectly fine until you arrive at that perfect distance between "too far" to turn back and return it, and "too close" to actually being able to begin placing things in the cart. With the Faker, once you pass that perfect distance, you begin realizing you should have taken it back and you're stuck the entire time wishing you had taken it back for a better cart. These are the relationships that after you have invested so much time into them you try to force them to work. The people you regret you chose and the boundaries you wish you had never crossed. Have you ever been in relationship and you've had to say to yourself, "This person is not who I thought they were?" Better yet, "I wish I had known him better." That's a classic line from those encountered by the Faker. Don't settle for the Faker.

Have you ever gone to the supermarket and encountered shopping carts that were defective? What did you do? How did you respond? How you responded is probably indicative of how you would respond in a dating relationship. I know it sounds preposterous but just think about it for a minute. I'm embarrassed to admit it but recently, I found myself grocery shopping with a warped cart. The problem was I couldn't seem to find a decent cart in the store. After so many failed attempts at trying to find a good one, I settled for the best of the worst carts. This is what we do in relationships everyday. We settle for broken, warped relationships because the last few we experienced failed to meet our expectations, so we lowered our expectations to meet our experiences.

When I go shopping now, I have three non-negotiables:
1. I need my headphones so that I can tune everybody out and dance up and down the aisles to my music.
2. I need a plan of execution.
3. I must have a fully operational shopping cart. I've learned to take my time and make sure that I have a cart that I am fully satisfied with.

If I can't find the right cart in a decent amount of time, I will scratch shopping altogether and choose an entirely different store in which to shop. I say to myself, "If I can't be fully satisfied with the right cart, I'm sure I won't have a good overall experience shopping."

Dating should be treated the same way. That is:
1. You ultimately have the power to choose who you date.
2. You don't have to settle for a man that is deficient, defective, and/or dysfunctional because you feel there aren't anymore good men left out there.
3. If you can't seem to find anybody that meets your satisfaction, consider the place in which you've been looking. Maybe you need to scratch dating altogether? Perhaps give it a rest and try again later.

    I remember Cicily. She was intelligent and very focused but after spending 3 weeks dating her, I realized she had only one thing on her mind - SEX. I've been there and done that and I have a shirt to show for it. I didn't need the distraction of having to constantly fight the temptation of sex and the bondage that comes along with it. I mean, I'm no virgin and I do practice abstinence but when the other person you're trying to date doesn't operate with the same conviction of being willing to spend time together outside of the bedroom, then you, my friend, have some decisions to make —— I just didn't need that drama in my life so I moved on from her.

    Vanessa was gorgeous and I had known her from high school. She understood me and we had history together, but there was one problem - she didn't care for children and certainly didn't want any of her own. I knew this was an issue but I teetered back and forth with it and tried to get myself to make it work - truthfully I knew it wasn't going to work. As beautiful as she was, I had to accept that she and I had a different value system.

Now Chaleece, I would have gone to bat for her. She was meek, accomplished, and she had a good head on her shoulders. I loved her family. The problem was that she was so broken and bitter from her past relationships that I found myself spending most of my time functioning as her therapist and the rest of my time trying to get her to trust me as a result of her past. There are some relationships that you will obviously be willing to go the extra mile and do what you never have done before, but you have to draw the line somewhere. You can't be a therapist, firefighter, and the boyfriend, too.

> NR
> You can't be a therapist, firefighter, and the boyfriend, too.

The point is, you don't have to settle when your values don't align and when you honestly feel you can do better. We go through life and through relationships with expectations of others that fail to be met. But most of the time, it's because we are in relationships with people that we honestly know we shouldn't have been with in the first place. Most cases of resentment, bitterness, and confusion can be avoided if we stop settling in the first place. There doesn't have to be an expectation gap filled with confusion and disappointment if you make a conscious decision that lowering your expectations is just not worth the heartache of settling for mediocrity. I'm sure these women will all make some man very happy someday, but the lesson for me is that every person that looks like a great candidate does not have to be YOUR candidate. You don't have to settle.

YOU'RE WORTH THE WAIT

I recently woke up to a message on my Facebook newsfeed that read, "Keep calm, I just got engaged!" Shortly after her rant about giving dating a break, Shelly posted the news about her newfound engagement. It had just been a matter of days when she told me the word "dating" should be expunged from the dictionary. We spoke candidly about the red flags in her relationship, and she had agreed it was time for her to move on with her life. I gave her some advice on how to get over her situation and I encouraged her to take time to focus on herself. A few days went by. As I scrolled through her news feed, there she was, advertising her new beautiful ring. I wanted to be happy for her but unlike everybody else, I knew more than what others knew. "I just want to be married," she would confess to me. She wanted to be married more than she wanted anything else. At that moment it struck me, marriage was her destination.

Earlier in the chapter we defined dating and redefined marriage as something more than a destination or a goal in life. The purpose of dating may be to determine one's suitability for marriage, but then what? You get married? Now what? Marriage is much greater than something to accomplish. It's much more than a wedding dress and wedding photos, and even building a family. Marriage is the calling of and the establishing of a ministry together for the glory of God. Marriage is primarily and solely for the benefit and glory of God, not us. If this is something that one or both people do not consider as they are pursuing marriage, they may soon have bigger problems on their hands.

Nearly 4 weeks later, the engagement was called off. Shelly was humiliated. She spent so much time publicizing and advertising her engagement ring and very little time analyzing the driving force behind her overbearing desire to say, yes. In our conversation she said, "I have learned my lesson. I just need to wait." Waiting is probably the most difficult thing to do when it comes to finding the right life partner. When I think about the ability to wait patiently despite time, age, and circumstances I think about Ruth. Ruth 3:18 says, "Then Naomi said to Ruth, 'just be patient, my daughter, until we hear what happens. The man won't rest until he has settled things today.'" Naomi reminded Ruth, You are worth the wait. You are still necessary. You are still valuable. So act like it.

Ruth, a widow, had lost her husband and in all likeliness given up on finding love, again. She traveled with her mother-in-law to a foreign town and made Naomi a commitment to look after her in Naomi's old age. One day while Ruth was out in the field and trying to earn enough food to care for Naomi, she caught the attention of Boaz, a prominent man from the town. After expressing an interest in Ruth and Ruth following the sound advice from Naomi on how to respectfully show her interest in a man, the best advice was not yet given. I imagine Ruth had become a bit flustered and impatient and wanted things to happen sooner than later. This is why we all need a Naomi in our lives to remind us that at the end of the day, it is better to wait and allow God's will to do what our hands cannot do. After some time, Boaz returned and made Ruth his wife. From this union and family, Jesus would eventually be born.

> Naomi reminded Ruth, You are worth the wait. You are still necessary. You are still valuable. So act like it.

Ruth teaches us three things:

1. Be fruitful not wasteful.

In this narrative, just like in the narrative of Adam and Eve, Ruth is busy doing what she feels deeply convicted to do. It's at this time that God chooses to bring the right man alongside her. She's not wasting her time complaining or focusing on her circumstances, she is walking in purpose and accepting God's will for her life. I know it is easier said than done but it doesn't mean it's not relevant and true. Sitting by and waiting for a man or a woman to fall from the sky while you're having a pity party isn't the best remedy for happiness and fulfillment. You have to achieve fulfillment before there is somebody else in the picture. What are you doing with your time? What are you doing with your life? Are you being fruitful or are you being wasteful?

2. Your Boaz comes with selectivity not impatience.

Ruth wasn't an old woman; she had the opportunity to choose younger more popular men to be connected to. Despite her many options, she was selective. She didn't need to date many men nor did she need to sleep around. When she finally encountered Boaz, who just happened to be an older man, she selected him. Even Boaz was surprised that Ruth was interested in him. Ruth didn't operate from a platform of impatience; she was careful and selective. When the time eventually came for the right man, Ruth wasn't used up. She wasn't holding on to her past mistakes or circumstances, she was ready. You can't be selective and impatient simultaneously.

3. When you know how much you're worth, you don't give people discounts.

What is the rush? Why enter into a relationship that is motivated out of some self-imposed timeframe? Why devalue your worth? Ruth didn't settle even after her first marriage eclipsed and despite the fact she didn't have any children yet. Ruth waited for Boaz, she didn't settle for anybody else. She knew that she had something to offer a man if the right one ever came around. She knew her value and she acted on it.

When Shelly's relationship didn't work out, her first message to me was, "Well, I know what you're going to say, 'I told you so.'" From her account, she had given him the benefit of the doubt. She knew there were unresolved issues in their relationship but she figured they would be able to work them out before they got married. She added, "I wish I had known then what I know now. I wish I had waited before I let him move in with me. I may have to be alone for the rest of my life but I would rather be alone than be with somebody who doesn't value me."

When you know how much you're worth, you don't give people discounts.

We believe that our dreams of wedding rings and honeymoons will solve all of our current problems without recognizing that they will simply create new variants of the same problems we are experiencing right now. If there are some things you need to address now, go ahead and address them instead of pursuing some ideal in the future. There is no need to rush. You are worth the wait. You are still necessary. You are still valuable. So act like it. It may have taken some time but eventually Ruth found her Boaz.

FIVE QUESTIONS TO ASK BEFORE RE-ENTERING THE DATING SCENE

After spending lots of time talking to women and men about dating, conducting my own research, and making my own observations, there are five important questions you should ask yourself before entering or choosing to re-enter the dating scene.

1. Why am I choosing to date?

This is probably one of the more unsuspecting and shocking questions I asked to many people that I counsel. Most of us don't really know how to answer the question because it seems so simple. When I ask the question I usually get a very awkward pause followed by stuttering and a fluff answer. So do me a favor, take a long hard look at what is driving you. Are you looking for friendship? Are you hopeful for marriage? Do you just want to play? Are you bored? Are you lonely? Are you seeking somebody to keep you company or to fill a void? Each of the answers to these questions will make a world of difference, especially for the person you are trying to date. Be crystal clear about your intentions. More importantly, be clear and honest with yourself about your intentions and motives for choosing to date.

2. Am I the best possible version of myself, today?

There is no reason for you not to give somebody the very best you. It doesn't matter who the person is, they deserve to meet the best you. Nobody deserves the responsibility of having to become your therapist, bandage you up, and make excuses for your past, just to be able to make your dating relationship work with them. It's your responsibility to become the best version of yourself before trying to create something with someone else. That's just like walking on a car lot and engaging a car salesman and after negotiating the terms of a new car, he goes into the back and drives around an old, beat up car to give you when you are expecting a brand-new presentable car – that's not what you signed up for. Nobody signed up to get an old, beat up version of you. You have to ask yourself, "Am I really ready?" Are you moving too fast? Are you carrying too much baggage with you? When was the last time you actually dated somebody? Has it been too soon? Has it been too long? These are important questions to ask yourself and how you answer will determine your best course of action.

3. Can I handle another unsuccessful relationship?

Do you know how often I am approached by somebody or receive an email from a person who has decided to leave my ministry or the church in general because a dating relationship didn't work out, which has left them distraught and disillusioned? Too often. As if they didn't realize that dating comes with its own share of risks, one of which, is that it may not work out, and if it doesn't, it should not mark the end of your ability to cope with life. People move from one unsuccessful relationship to another relationship with little or no emotional capacity to pay the price of what it costs to be in a healthy relationship. I'd like to think that dating is similar to using a debit card. If you know that you have very little money in your checking account, you might not want to be at the store trying to shop for more than what you can afford. I can go from one store spending

> **NR**
> Nobody deserves the responsibility of having to become your therapist, bandage you up, and make excuses for your past, just to be able to make your dating relationship work with them.

money to another store and spend more money without replenishing money into my account. That's why so many people are in debt now. Can you handle another unsuccessful relationship? Well, it depends on how much you have in your emotional checking account.

4. Do I have my own spiritual walk?

Dating relationships will test the existence and the durability of your spiritual life. Whether we like it or not, dating can serve as a distraction or a catalyst for growth. Oftentimes dating negatively impacts our spiritual life because it divides our attention and our devotion to God. Apostle Paul tries to forewarn believers at Corinth to consider remaining single and celibate because to date and to entangle oneself with another person means your attention would have to be divided; and as a result, your spiritual life would likely suffer. It is of the utmost importance to not only have your own spiritual walk but to be actively growing spiritually. It is also important that you take the time to assess whether you are ready, spiritually, to invite romance into your spiritual life.

Although there is no barometer to determine a person's spiritual readiness for a relationship, there are some questions you can answer that will help to assess the fragility of your spiritual walk.

Do you have your own personal and consistent devotional time? What does it consist of? Can you articulate your personal faith story? How often do you pray? How often do you seek God before you make an important decision? What are your spiritual struggles? What are your spiritual gifts? How often do you use your spiritual gifts and talents? What does your faith mean to you?

Based upon your answers, do you feel secure enough to invite romance into your spiritual life? It is essential that you have your own spiritual walk so that you won't depend on someone else's.

5. Am I genuinely happy? Am I fulfilled without a relationship?

Yeah, yeah, yeah, I know. You're happy all by yourself. I have heard it a thousand times. Social media is always trending with people claiming they are happy and fulfilled being single. So many people vie to keep up the facade that they are happy all by themselves, and they don't need to date. I often hear, "I'm content with being single, I don't need a man." People have learned to trick themselves into believing that they are truly happy when, for many people, it's just a cover up. The best remedy to test your theory is to take out a sheet of paper. Fold it in half. On one side, put at the very top "How many dates have I been on in the last 10 months?" On the other side, put at the very top, "how many relationships have I been in in the last 24 months?" This should help to qualify your statement. If within the last 10 months you've been in more than one relationship, or within the last 6 months, you've had more than a hand full of dates, I question whether you are fulfilled living a solo life.

CHAPTER 2

THE TEMPTATION OF IMPATIENCE

OVERCOMING OUR HIGH DEMAND FOR MICROWAVED RELATIONSHIPS

"I hold it true, whate'er befall; I feel it, when I sorrow most; 'Tis better to have loved and lost than never to have loved at all." - Alfred Tennyson.

These words are so beautiful and yet so misleading, in my opinion. The idea of achieving love by any means necessary is not only destructive, it's dangerous. I once heard somebody say, "The best way to approach love is to believe everything and everybody is a source of joy and enrichment. Then face life chest on, receive your blows of pain and betrayal, and survive until you find everlasting love." If this were true, I might as well leave my front door wide open at night and go to bed and hope that I don't invite any unwarranted guests into my home while I sleep. My point is that living this way wouldn't be a great idea.

The heart is both sacred and fragile and we should not carelessly sell it to the highest bidder nor should we give our heart to any willing candidate, for the sake of being able to say we once experienced love.

My Bible teaches me that I have a responsibility "to guard my heart above all else" (Proverbs 4:23), because "unrelenting disappointment will leave you heartsick" (Proverbs 13:12). Many times, our overbearing desire to love and to be loved and to have companionship causes us to make hasty decisions and to operate from a spirit of impatience. As a result, we leave our hearts open and unguarded which often leaves us broken and disappointed and wallowing in our regrets. "I wish I had waited to get married." "If I could do things over again, I would not have rushed." These are voices I replay in my head from people close to me who were robbed by impatience. The heart is both sacred and fragile and we should not carelessly sell it to the highest bidder nor should we give our heart to any willing candidate, for the sake of being able to say we once experienced love. Our chief goal isn't to "experience" love as if it is something to remove from our checklist, but our desire should be to cultivate lasting relationships that please God. This begins with patience, seeking God's will, and developing the right mindset.

"JUST ADD WATER" TO YOUR RELATIONSHIP

We live in a time when immediate gratification and quick results are coveted by the masses. If something requires a process or if it means you will have to wait, you can expect that your idea will almost always be unpopular. We are spoiled by convenience and we are taught to be dissatisfied with the idea of "later." We obsess over the word, "instant" and we are so enamored with the idea of "right now" that we don't spend time to cook and we spend even less time to eat. In comparison to the rest of the world, Americans spend little time in food prep and dining, yet our obesity rate is nearly double that of other countries. The average time spent cooking an evening meal is now less than 27 minutes in comparison to spending over a full hour back in the 1980s. According to the Organization for Economic Cooperation and Development, we cook the least and eat the fastest. A weeknight meal can become so preplanned and manufactured that it gets squeezed into that narrow time slot between your getting home from work and your desperate need to harass your smartphones for your evening fix of daily news gossip provided by social media, right before taking your shower for bed. But wait! Not before you miss your favorite television show that you recorded on your DVR. Hundreds of millions of dollars are spent on instant meals each year. Once upon a time, a piping hot, freshly cooked meal was an essential ingredient of our daily lives. Now our meals become so processed, frozen, dehydrated, and prepackaged for our convenience that all we have to do is, "Add Water" or "Microwave It" and there you have it, an instant meal. Unhealthy, albeit, but still instant. The price we pay for convenience. We just grab dehydrated food kits off store shelves and add water. In the simplest of terms, we cook fast, we eat fast, we become obese, and we die early. Good grief. Well, what does this have to do with relationships, you ask?

Healthy relationships do not come in a jar, nor do they come prepackaged.

After screwing up many relationships, I think the most important concept I've learned is that I cannot dehydrate a relationship of the essential things it needs simply because I desire immediate results. I've learned that there are some things in relationships that I can't manufacture, borrow from anybody else, or expect to come "ready-made" to work. Healthy relationships do not come in a jar, nor do they come prepackaged. When it comes to the temptation of being impatient, one of the basic attitudes we fall victim to, is the mindset that we can just "add water" to relationships and expect for them to work. We have become obsessed over "instant" relationships - we exert little effort and expect for our relationship to thrive. We move fast, we commit blindly, and we love hastily—yet we wonder why the cycle of dysfunction cannot be broken. We pursue relationships too hard. We fall too quickly, and we divorce too often. Then to make matters worse, after our breakup, we sprint to recovery and/or we get upset with God if our heart doesn't heal immediately.

It takes time and intentionality to develop the right mindset and to establish the proper values in order to foster a healthy relationship. This is what the temptation of impatience explores. If you are choosing to date, your commitment as a believer is to learn to develop trust, character, and true intimacy in your relationship. This requires an investment of time and energy on your part, and a willingness to put God's will before your desires. But like all other aspects of a healthy relationship, taking the time to develop trust, character, and intimacy pays high dividends.

During my first year of becoming pastor of my current ministry, I met a beautiful, charismatic, and intellectually driven woman named Soya-Renee. She is the kind of person men drop dead for. I mean no man in his right mind would hesitate to make her his wife. She is magnetic and has a great personality. When I first met her, my immediate reaction was to recycle my behavior to be impulsive. I was still single with no family so I instinctively wanted to make her my candidate for marriage. I was single, she was single, and we both were attracted to each other, so why not pursue a relationship together, right? Why not ring the wedding bells? In my mind I had already hijacked her name and welded my last name with hers and it seemed to fit perfectly. This was my process, although it appeared that God was trying to show me a better way to pursue healthy relationships other than what I had known.

Both Soya and I met each other during a time in our lives where we were looking for something different. I wanted to try a different approach to relationships and she also wanted to do things differently than before, and we were both spiritually mature enough to know that if we were going to get anything out of our friendship, we could not cross certain boundaries. We had to be intentional about doing things differently. We wanted to seek God's will, which meant if God's purpose was for us to only be friends, then we wanted to honor that. There was no sex, no kissing, and no touching. I know, it sounds impractical and boring, but it wasn't. It was very fulfilling and rewarding. The time we spent together was intentional and had healthy boundaries. We were committed to being intimate but only through self-revelation, vulnerability, and transparency, not sex or anything alluding to it. There were no representatives allowed in our friendship. That is, when we meet people we often waste so much time introducing them to the people we want them to think we are, and even who we think they want us to be rather than just revealing to them our truest "self." Truthfully, I think we hide and pretend so much that many of us couldn't identify our true selves in a police lineup if we were paid to do so.

True Intimacy is developed over time, and there's no way to microwave it.

We were committed to doing things the right way. As a result, she taught me intimacy. We learned more about one another in 21 days than I've ever learned about any woman over the course of six months. I simply risked being patient, and I took a chance in learning to become more vulnerable. Nearly 2 years have gone by and we still casually converse and engage in stimulating intellectual and spiritual conversations whenever time permits. Trust. Character. True Intimacy is developed over time, and there's no way to microwave it. You can't dehydrate it and expect to just "add water" and voila, a healthy relationship! Your commitment is to say, "over time, as we are becoming friends, I want to learn you in ways nobody has ever done before and I want to reveal more of myself to you. I want to risk becoming intimate with you, which is why I want us to risk sharing more of who we are without preoccupying ourselves with bedroom tango."

Soya-Renee and I are not dating, but we are great friends. There are no regrets because we chose not to dehydrate our relationship of essential ingredients in order to experience instant gratification. We chose not to "just add water" to get the quick results like many have chosen to do, or like I have chosen to do in the past. Healthy relationships take time. They require you to think independent of and contrary to the popular opinion. Healthy relationships don't just happen; you must create a space for them to thrive or else they will plummet. You have to be patient enough to seek God's will which may sometimes mean the person you think is supposed to be your wife or husband actually was just meant to be in your life for a season or for a reason. Don't corrupt it because you chose to give in to the temptation of being impatient.

CONVENIENT or COMMITTED?

So many people are involved in relationships that are being held together with the glue of convenience rather than the promise of commitment. The same quick grits and powdered egg breakfast commercials that tell us convenience is better, also reinforce the idea that the traditional way of doing things is old and outdated. We become so preoccupied with trying to find a qualified candidate to be in a relationship with that we fail to realize the difference between being happy and just being plain ole' distracted. Like my close relative who has been diagnosed with Attention Deficit Disorder (ADD) and is easily distracted by irrelevant sights and sounds and quickly bounces from one activity to another, we behave similarly in "convenient" relationships. We need something to do with the available time on our hands so we distract ourselves with convenient people and empty connections that don't really accomplish any significant purpose other than the superficiality of right now. A convenient relationship only feeds our needs and fills our voids for today.. Whether the void is sexual, emotional, or financial, we entangle ourselves in these relationships out of convenience. You're in a convenient relationship because you've diluted the connection between you and the other person and you've chosen to dumb down your expectations in order for the relationship to remain afloat.

There is a difference between being in a relationship and just being together because that's all you're used to. Are you in a relationship or is it just a routine? Are you truly happy or is your relationship just a distraction for you, or better yet, for the other person? Are you there just until something or someone better comes along? Is your relationship convenient or does it have the promise of commitment? There are five signs to help you determine if you're just a convenient distraction for him or her.

1. There is no real commitment.
Where do you stand? She says she loves you and that she really enjoys your company. She thinks you are a really nice man. It has been nearly three years now and you have received yet another recycled version of "I'm not sure." You have her hand but you don't fully have her heart.

If you truly had her heart she would commit to you and allow all other remnants of doubt, fear, and procrastination to fall by the wayside.

2. You don't feel like you can truly be yourself.
Can you speak your mind freely without fearing ridicule? If you can't be your true self, don't blame him although he may deserve some of the blame. You need to hold yourself accountable. If you can't be yourself, you should attempt to address this in your relationship. If you can't resolve this matter, do not ignore it. This is something that most people wish they heeded long before they got unhappily married.

3. You're not having fun.
Is the routine killing you? Are you moving any closer to your goals and aspirations as a couple? Is he challenging you to be better? Do you have genuine fun together or does it feel like you're walking in quicksand? Do you both make a conscious effort to focus on one another outside of your jobs, daily obligations, and your weekly responsibilities? Do you truly enjoy the relationship or do you tolerate it? Do you spend most of your time fighting or navigating through negative energy in your relationship? It's truly hard to love someone when you are not enjoying him or her.

4. You don't talk about a future, together.
Why don't you ever talk about children together? It's probably because neither of you can see that future, together. If you want to take the pulse of your dating relationship, listen to the heartbeat of your conversations. Is it convenient for him or her? Well, beyond asking them for a simple answer, you can also pay attention to the

trajectory of dialogue. What does your relationship look like beyond today, tomorrow or even next year? Are you considering long-term and longevity? Yes, of course neither of you can see into the future but that doesn't mean you can't plan for it and talk about it and dream about a future that includes the other person as a main character in it.

5. You spend a lot of time alone together, but not around his family and friends.

If you have been dating for an extended period of time and still have not met his family or you don't spend any time around his inner circle of friends, there is a good chance he doesn't perceive your relationship as something serious or as having the potential for a future. Beyond how he may communicate his adoration for you, a man who is proud of the woman he is dating will gladly introduce her to the people who are closest to him. The same goes for a woman. This may be a subtle but revealing red flag. If this is a concern for you, it is your responsibility to address this as a matter of concern for you.

IMAGINARY ENEMIES

It is such a bad habit for me. Many mornings, I wake up and turn over in my bed, and while simultaneously thanking God for another day, I am checking my email and reviewing other's status updates through my newsfeed. I know who's got a new job, who is getting married, and where he purchased the engagement ring. I know who's newly single, who's having trouble being single, and who's acting like they are single. Within the first five minutes of my morning, I can see who's happy, who's sad, and who's indifferent. I can see who's traveling the world, who's following their dreams, who's starting their own business, and who's driving a brand-new car. Photos, status updates, and tweets flood my inbox and newsfeed every single morning, 24/7. And this is the consequence of always feeling the need to be "connected." I blame this on the technological age in which we live. My challenge, however, is that it has become so difficult to keep from looking at what is going on in somebody else's life and somehow refrain from comparing it to what is going on in my life. Sometimes I've easily allowed what I see happening in the lives of colleagues and distant friends to make me feel inadequate. The need to always be connected and the influx of information made available at our fingertips makes it nearly impossible for us not to compare our lives with other people.

> **NR**
> When we compare ourselves to others we are unconsciously overlooking the very uniqueness of God's imprint on our lives.

When we compare ourselves to others we are unconsciously overlooking the very uniqueness of God's imprint on our lives. When we compare our lives to others' we are in essence telling God we don't approve of his decisions, and that we believe we can direct our lives better than He can. When we become overly preoccupied with our daily newsfeed, we begin to develop what I call, "Imaginary enemies," or counterfeit truths based on superficial data. Imaginary enemies are obstacles, pitfalls, and self-imposed barriers that

> **NR**
> Imaginary enemies are obstacles, pitfalls, and self-imposed barriers that we erect and create for our lives based on false things we choose to believe are true.

we erect and create for our lives based on false things we choose to believe are true. We look at others' lives and then we tell ourselves, "If I don't get married soon, it's going to throw off the plans that I have for my life." Better yet, we may look at our friends' new profile picture and then tell ourselves, "I'm too old to still be single." I remember when I discovered that a colleague of mine purchased a new house with his new wife. (Of course, he did not purchase her, they were newlyweds.) After looking at his photos and timeline, I remember feeling as though I had nothing to show for the things I had accomplished in my own life. I easily disregarded my college and graduate school degrees, my international travel and missionary experiences, and the many lives I had empowered through God's word. I saw what I wanted to see and began fighting battles that never really existed.

As another temptation of being impatient, we create imaginary enemies by giving in to the dangerous habit of comparing our lives to other people's lives and to our culture's standards. When other people's successes and accomplishments become our vices, we have begun fighting an enemy within our own minds, that doesn't really exist. My close friend Pat told me she was giving social media a break. In her transparency, she confessed that when she was on Facebook, she would find herself examining her friends' profiles and becoming depressed. She decided to shut down all her accounts and fast from social media. Honestly, I thought she was just going through a phase, but after a while I began to see a totally different person emerge from this situation. Later, Pat expressed that she had become so consumed with analyzing the profiles of other people that she worked her way into a destructive mindset of regret, jealousy, and discontentment. In refocusing the energy she put into comparing her life to her friends' lives, she began developing hobbies, learning how to cook, and doing things she never really had time to do. Go figure. We spend so much of our time fighting unnecessary battles creating imaginary enemies.

"DON'T COMPARE YOUR CHAPTER 1 TO SOMEONE ELSE'S CHAPTER 20."

When I came across this quote, it screamed at me. "Don't compare your chapter 1 to someone else's chapter 20." This quote immediately gave me a release. It needed no explanation; it was obvious to me what it meant. I had grown into a bad habit of comparing my chapter to someone else's chapter. For example, within the past 18 months I decided to go back to the gym, but this time I wanted to take it to another level. I have a close friend who is a bodybuilder and trainer so I asked him to teach me how to properly develop my muscles and to become more physically fit. He put me on a regimen that would last eight months before I would arrive at a noticeable difference. Let me preface my statement by saying this; the gymnasium is a very intimidating place, but the gym is filled with people that focus more on what other people look like than focusing on their own individual self.

For the first half of my season training, I became discouraged every single day I entered the gym. Why? I was one of those individuals who silently looked at other people and then I would look at myself and immediately feel like giving up because I felt like I could never arrive at the place that they were. Their muscles were developed and all the people would watch them as they worked out. Half of the people in the gym would try to copy whatever they were doing. I nearly stopped going to the gym because I felt like I was wasting my time. My trainer reminded me that everybody has to start somewhere and that where I was, he used to be. This taught me a simple lesson that while we are so busy looking at somebody else's Chapter 20, we must understand that they too had a chapter 1. When we look at other people's lives, we don't know the story that lies behind their photos and we don't know what they had to endure. Their narrative is not our narrative.

I had a young man approach me in the gym and ask me how I got so physically fit and muscular. He appeared nervous and discouraged as he talked to me. It was if I were the epitome of the goals he was trying to accomplish but they seemed impossible to reach. I encouraged him. I shared with him the difficulties I experienced, and the times I felt like giving up. I shared with him that I had to learn how not to focus on other people in the gym but to focus on myself. I told him that I used to be where he was and that everybody has to begin somewhere. He was encouraged because from his vantage point, he would have never thought I had struggles of my own that I had to work through. This goes to show that when we compare our lives to other people, it becomes easy to make mountains out of molehills and to overlook the beauty of our own journey.

SEEDS BECOME TREES IF THEY ARE CULTIVATED LONG ENOUGH

"The mustard seed is one of the smallest seeds. However, when it has grown, it is taller than the garden plants. It becomes a tree that is large enough for birds to nest in its branches" (Matthew 13:32). When we allow a thought of inadequacy, failure, doubt, or that we are not good enough, to enter our minds by way of comparing ourselves to other people, it begins as a seed. Like any seed, it seeks to develop a root system and from that root system it will seek to mature and to produce fruit. When we learn to unconsciously disengage ourselves from the blessings that God has given us to crave the blessings that God is giving someone else, we are planting seeds of discontentment.

When we intentionally comb through our status updates and approach our relationships with the intent of becoming what we see in other people, our seeds develop a root system. When we experience disappointment and when our lives and our relationships don't turn out the way that we hoped they would, those seeds that have now developed a root system begin to produce fruit. Anger, envy, bitterness, depression, and many other fruits begin to grow from the seed that you cultivated.

When it comes to our tendency to develop imaginary enemies you must understand that seeds will become trees if you cultivate them long enough. Stop creating enemies that don't even exist. Stop allowing the seeds of doubt, impatience, self-worth, and negativity to pervade your heart and spoil your blessings. God is doing a great work in your life right now. Embrace where you are and what God is doing with you. Reverse the destructive habit of fighting and trying to be where God has somebody else. You are right where God wants you to be.

I WANT THAT

Currently, I don't have any children of my own. Therefore, I have willfully given my heart and time to my niece who has just turned two years old. She is now at the age where she thinks she can talk clearly but really she only understands herself while everybody else scrupulously tries to break the code of her translation. There are four words that she has learned to speak very clearly, "I want that" and she has learned the artful skill of coupling that phrase with the magic word, "please," which almost always seals the deal for her. It was so adorable when I first heard these words. She would lift up her tiny hands and smile with that pretty smile and those brand new teeth; "I want that….Peeesh." I mean, how could you say, no?"

Section One: THE NEW NORMAL

I remember one day we had a family gathering at my mother's home, and amid all the people and organized chaos in the house, there was this little girl moving about from one person to another practicing and carefully executing these four words and as a result, she was accumulating enough junk to open her own flea market. The things she was asking for I don't think she really wanted, or maybe she thought she did, but she really didn't know that what she was asking for perhaps wasn't good for her or what she really needed.

At that very moment, my niece became my teacher. She showed me myself in a dimly lit mirror. It has become particularly easy for me to move from one relationship to another relationship, from one chapter in my life to another chapter, holding hostage four words, "I want that. Please." I have learned the skill of holding out my hands to God and pointing my finger at different people, different relationships, and different marriages, and saying to God, "I want that. Please." Not realizing the things I've asked God for may very well look good to me, but they may not have been good for me. I've held my hands out and pointed my fingers at friends that appear to have what I want and I've said to God with no shame or hesitation, "That's the kind of woman I want, God. Just like Stephen. His wife is nice and humble and she pays attention to him." I've said this countless times. I've also prayed the prayers, "Lord, give me a woman that isn't combative and that would challenge me to be better. Let it be kind of similar to that relationship over there." Yes, that is until God would give me exactly what I thought I wanted, then I was forced to try to press the backspace button to modify them and even the Control, Alt, Del to try and erase those prayers.

I didn't know that when I was praying for a woman that challenges me, I was not necessarily taking into consideration that we may not share the same values about finances and about raising a family. And the cycle continues. Have you ever asked God for something only to be given exactly what you asked for and when you got it, it was not what you thought it would be? Better yet, you then realized you needed something more or different? We are a society ruled by perception and the truth is we are programmed to want what we see. This is what my problem has been for so long. It's a terrible temptation to think you want what you see. I had to realize that I was in love with the result, I wasn't in love with paying the cost.

When it comes to the temptation of impatience, the "I want that" attitude is one of the more inconspicuous yet destructive behaviors to possess. It's the habit of selfish will. Tunnel vision, should you call it. This is similar to Samson in the book of Judges 14:1-2. It says, "Samson went down to Timnah and saw there a young Philistine woman. When he returned, he said to his father and mother, 'I have seen a Philistine woman in Timnah; now get her for me as a wife.'" Samson saw a woman who caught his eye and immediately he wanted what he had seen. Nevertheless, before you can make it half way through the next chapter, this woman got married to his best man. The ultimate betrayal. From that point in his life, everything went downhill. What Samson thought he wanted wasn't what he needed and it wasn't what was best for him. There is an element of impatience and impulsivity that we must wrestle against

> **NR**
> When it comes to the temptation of impatience, the "I want that" attitude is one of the more inconspicuous yet destructive behaviors to possess. It's the habit of selfish will.

when it comes to dating and desiring microwaved relationships.

Sometimes my desire for companionship and to be with a woman to call my wife has been so strong that I, too, would act impulsively, think irrationally, and make decisions based solely on, "curb appeal," or the attractiveness of the exterior, just like Samson. Embarrassingly, there are times I've seen women who pleased my eyes, who were very articulate, that I would immediately lose focus and begin pursuing the imagination of merging our last names (such a bad habit). Sometimes I have been just like a consumer that was looking for a house, but was primarily focused on curb appeal. Of course, I understand that just because a house has good curb appeal doesn't mean the house is in good condition. It's just made to appear to be in good condition.

Don't fall in love with who you want the person to be, fall in love with who that person is.

Samson helps us to better navigate through our temptations to be impatient by teaching us three things:

1. Every thing that looks good to you isn't necessarily good for you.

2. Don't fall in love with who you want the person to be, fall in love with who that person is. In order to do this, you must take your time. There is absolutely no need to rush.

3. Don't forget to take the time and consult with God before jumping into any relationship. (Don't forget to take the time and include God into your equation for marriage).

Maybe, just maybe, had Samson taken his hands off the mute button and actually listened to the wisdom and counsel of his parents and spent time seeking God, he would not have made the impulsive decision to quickly get married. There are some instances of 'love at first sight' but for the rest of the world, time can be the deciding factor. Taking your time and seeking God's face and allowing room for your emotions to settle will help to keep you from making the greatest mistake of your life.

Remember that although you think you would be better off with somebody else's life or you may want to be married to a person like your friend's spouse; you are simply desiring surface level things. You may have said to yourself, "That's what I want" or "Why can't I have that?" This is typically a reaction to what we see with our eyes and hear with our ears. We easily fall in love with the result - the image of satisfaction, happiness, marital bliss, contentment, but we aren't too fond of loving the process. We must first learn to embrace and appreciate the life that we have now, and work through our flaws, embrace our limitations, and maximize the gift of singleness that God has orchestrated for our lives. The truth is, I've come to learn that aside from the guidance of the Holy Spirit, we have no earthly idea what we want. While reading, I came across the saying, *"Thank God for protecting me from what I thought I wanted and blessing me with what I didn't know I needed."* This rings true on so many levels, especially in relationships. We can create lists and values and job descriptions of potential employees but we really don't know what's best for us. I have learned this to be true.

> **NR**
> Taking your time and seeking God's face and allowing room for your emotions to settle will help to keep you from making the greatest mistake of your life.

CHAPTER 3

DATING YOUR INSECURITIES

RECOGNIZING THE DAMAGE YOUR INSECURITIES CAN CAUSE TO YOUR DATING LIFE

Why didn't it work, you ask? My engagement? When I view the topography of our relationship, it was mostly me. More specifically, it was my insecurities. I heard a prominent pastor, James MacDonald, describe an insecurity as "the awareness of the gap between who you are and who you want to be." I imagine a large room where there are two chairs opposite of another, each facing forward. One representing where I am now and the other designating where I want to be. Imagine the deafening gap in between the two chairs to be signified by a body of water. For me, the embarrassing acknowledgement is that I have spent so much of my life in one chair trying to get to the other chair, but the more I try, the more I feel like I'm drowning. I've spent most of my life treading water and swimming in the gap.

After recently completing my graduate degree from Duke, I failed to land a lucrative job. I wasn't able to provide for her, financially. I wasn't where I felt I needed to be to make her happy. I literally measured my feet and realized I couldn't fill the ginormous shoes of her father. He was a man's man. Her mother was such a blessing to me. A kind spirit. My family didn't look like her family. They were organized. Achieving. Together. Supportive. All things I had wished my family was. It was hard. It wasn't their fault, it was mine. I was hypersensitive about my family and the poverty from which I came.

When I looked at where I had come from, my mother and father didn't even finish high school. We never owned a home, furthermore, an automobile. There were no savings in our accounts, because we had no accounts. I was only introduced to a checking account at the bank when I went to graduate school. We had no investments. We knew nothing but how to live from paycheck to paycheck. Although her family had to pull themselves up by their bootstraps at some point, they were at a different place. More like worlds apart. Ultimately, it came down to the fact, I wasn't confident that I could provide for her the lifestyle she deserved. I had grown so accustomed to struggling that I allowed my insecurities to rob me and manipulate me into believing lies about who I was and the husband I was incapable of becoming.

> **NR**
> Insecurities come in all shapes and sizes. They disguise themselves and worm their way into our lives and we do not take notice of their births.

Insecurities come in all shapes and sizes. They disguise themselves and worm their way into our lives and we do not take notice of their births. They grow right before us, befriending us. We nurture them as our own children. We become protective over them, and like a teenager reaching puberty, our insecurities give us heartache and grief. They betray us at the moment we least expect them to. With my insecurities, I kinda felt like Gideon. God selected him to do something great but Gideon's family background and where he had come from

> **NR**
> Insecurities are those things we learn to hide and dress up to pretend they don't exist, but most times, they are the things that sabotage us and keep us from reaching our true potential.

became a point of insecurity and frailty for him. "But God, how can I rescue Israel? My clan is the weakest in the whole tribe of Manasseh, and I am the poorest in my family" (Judges 6:15).

Insecurities are those things we learn to hide and dress up to pretend they don't exist, but most times, they are the things that sabotage us and keep us from reaching our true potential. Unfortunately, my insecurities became my fatal distraction. A cycle I am restlessly trying to break.

NO REPRESENTATIVES, PLEASE

He was overcompensating for what he did not have and for who he was not. Yet, Priscilla had no clue. He often told her, "I know I am not the man you deserve but if you give me time, I will become that man." In the moment of romance and ignorant bliss, Priscilla thought, "He is so sweet. I don't deserve such a kind and thoughtful man." You see, what Priscilla failed to accept was that Tim wanted to be with her so badly that he was willing to do whatever he needed to do to get her to say, yes. He had been pursuing her relentlessly for about 18 months but she declined his offer each time. He wasn't her "type" and he was still a

Substitution by suffocation happens when we try to suppress and hide our insecurities so people can see the part of us we want them to adore.

little rough around the edges. Eventually Priscilla thought maybe she was being blinded to a blessing that God was trying to give her, so she gave Tim a chance. The thing that nobody realized was that Tim secretly felt he did not weigh up to her standards, so naturally he did what so many of us tend to do: he hired a representative to perform for him in their relationship. He fought to be who he thought she wanted him to be and dismissed the person he really was. It's called substitution by suffocation. Substitution by suffocation happens when we try to suppress and hide our insecurities so people can see the part of us we want them to adore. I was watching the Conan O'Brian show once when a celebrity confessed she didn't like new relationships because she felt that she had to be a fake version of herself. Therefore she learned the art of sweeping her crazy under the rug for the first few months.

I digress.

Every week, Tim sent Priscilla flowers to her job and to her home. He mailed her Hallmark cards so that she could read them in her free time. He often expressed to her that he would work to become the man that she wanted him to be. As a result, she felt the need to write him daily letters to affirm, validate, and encourage him and to assure him that he was "enough." Priscilla thought this was a normal thing in order to be considered a good wife, but I thought otherwise as she stood before me reminiscing on her newfound love. Yes, we all should be encouragers, but you shouldn't feel obligated to boost and validate a person in order for your relationship to thrive. It becomes more of a job than a pleasure.

The disconnect happened when Tim looked at his past and compared it to Priscilla's. It forced his insecurities to surface and then hijack the relationship. Priscilla wanted to date and marry so badly that as soon as she found a candidate she made him "the one." She told me, "I never knew being in love felt like this. It's magical. He buys me flowers every day. He brings me cards. He takes me out to eat. This is so amazing. He values me." Although Priscilla was an early thirty years of age, she didn't have much experience with romantic relationships, especially not my type of experience. I didn't want to rain on her parade as people often say I do, but I felt she was in love with the "idea of love" rather than being in love with the person. Two days later, she discovered inconsistencies and infidelities in their relationship. In retrospect, as she sat across the table from me trying to radar the collapse of her relationship, Priscilla reflected on the moments she thought were romantic but now realized they were clues that she was really dating the representative. "Man, he really knew how to act," Priscilla shared with me as she unveiled the many bones that came falling over from his closet. His infidelity and his blatant lies were something she never would have guessed would have happened to her and when she confronted him, he looked her square in the eyes and said, "I am the man you fell in love with." Actually, she fell in love with your representative, Tim.

There is a cycle in relationships that I have come to notice. We meet people and we often find ourselves on one side of the spectrum; either we feel we are not good enough for the other person or the other person feels they are not good enough for us. In any case, we often engage in relationships where there are representatives who spend more time with our significant other than we do ourselves. Insecurities make us feel that we need to overcompensate in relationships just to get the other person to receive us. Insecurities rob us of the freedom God is calling us to live in. We allow our insecurities to hijack our lives when we hire representatives to perform in our relationships and for the people we meet. We hide who we really are and try to become somebody we are not just to impress somebody who probably isn't worth impressing

> **NR**
> Insecurities rob us of the freedom God is calling us to live in.

in the first place. We become excessively compliant in relationships, overly spiritual, hyper-talented, and the list goes on. We hire these representatives primarily because our insecurities dictate to us that wearing a mask and hiding our truest self will cause the relationship to reach its greatest good when that couldn't be furthest from the truth. Representatives hurt relationships rather than help them. You cannot truly hide your insecurities, you can only suppress them for a period of time.

THE BIRTH OF OUR INSECURITIES

Whenever I do a seminar, I typically begin by asking my audience a simple yet essential question: what is God's purpose for your life? What has God called you to do? Believe it or not, I am almost always met with a very awkward silence and a few people who try to justify why they have no clue how to answer the question. Why do I ask this question? That's because most of us spend more time trying to force fit glass slippers on the wrong person than trying to align our lives with God's will. I ask this question because it helps people to realize that there are some things that are more important than just pursuing a dating relationship. Who are you? What is God currently doing in your life? If you can't answer these things, then these things won't become a priority when you start dating. I have found that the breeding ground for insecure behavior is when we are not certain about who we are and our essential life's purpose.

> **NR**
> Our desire to be something and somebody that we are not can be so ingrained in our psyche that many times we cannot see our present partner clearly—we see them as a savior, a rescuer, or a shadow to hide behind.

Insecurities happen when our expectations and our reality don't intersect. Our way of coping is carried from relationship to relationship, from decade to decade. We avoid taking a hard look at ourselves and dealing with the things that sabotage our relationships. Our desire to be something and somebody that we are not can be so ingrained in our psyche that many times we cannot see our present partner clearly—we see them as a savior, a rescuer, or a shadow to hide behind.

We become bad stewards because very little time is spent on ourselves while we become prodigals to other people's expectations. We waste most of our time revolving our world around other people, meeting and satisfying other people's needs. We make excuses so we don't have to think long and hard about our own lives and what we desire from it, and more importantly, how we plan to get there. This is the breeding ground for insecurities. This is when the conception begins. Defensiveness. Detachment. The need for constant affirmation. Always projecting our dissatisfaction and discontentment onto others — all are byproducts of the insecurities we nurture in our lives. We cling mostly to people who make us feel comfortable and who feed our voids rather than reveal them or challenge us to deal with our own deficits.

But what happens when we engage or partner in relationships with confident and secure people?

Two things. We either challenge ourselves to grow and to become better, not at the expense of the other person, or we find ourselves overcompensating and setting ablaze a trail of fears and anxiety that disrupts the chemistry and organic growth of the relationship.

In these relationships, we find ourselves making excuses and trying to be better at things we were never really good at doing. We allow the other person's confidence and assurance of themselves to cause us to act out and feel threatened. Their accomplishments become like mirrors that force us to notice the things we have failed to accomplish. Their security heightens the awareness of our own insecurity. We create tension instead of fostering trust and we wage war with our partner as if they were our enemy. It's like an insecurities exposé. This is the origin of our relationship's demise. We try to romance our insecurities and dress up our fears because we secretly feel we don't measure up to the other person, somebody that is likely not God's send to you anyway. This is how our insecurities are born, how our avoidance to discover or to rediscover our identity finally finds its chance to implode, how our happiness is disrupted and detoured. Insecurities are born and we often force the people we love to serve as our OB/GYN physicians to reluctantly give birth to them.

CARRY YOUR OWN BURDENS

I can tell you a story about how we all live happily ever after. I can tell you a story that if you read your Bible and jot down a few notes about self worth, voila, your insecurities will vanish. But that would be a lie. You can crowd your exterior by reading many self-help books and fill your life with a busy schedule so you don't have to face yourself. At some point you're going to have to face who you really are - a boy inside a manly body hoping to one day find somebody to affirm your insecurities. These are the words I have had to speak to myself in many seasons of my life. The truth is, other people aren't the problem. The scary part is coming face to face with who you really are. I'm just like many of you, I have weaknesses. I have insecurities. I wrestle with my self-worth and whether I will ever reach my full potential in life. I have an overwhelming desire to someday have a wife and children to call God's gift to me but sometimes those desires become an insecurity for me.

It's downright exhausting to carry around my own insecurities, but it's debilitating when you have to manage other people's insecurities, especially in relationships. Nobody deserves the dreadful responsibility of having to be a chauffeur for your insecurities. Even God got frustrated with Moses as he tried to help him alleviate his insecurities and reassure him that he could thrive in spite of them.

> It's downright exhausting to carry around my own insecurities, but it's debilitating when you have to manage other people's insecurities, especially in relationships.

10 But Moses pleaded with the LORD, "O Lord, I'm not very good with words. I never have been, and I'm not now, even though you have spoken to me. I get tongue-tied, and my words get tangled." 11 Then the LORD asked Moses, "Who makes a person's mouth? Who decides whether people speak or do not speak, hear or do not hear, see or do not see? Is it not I, the LORD? 12 Now go! I will be with you as you speak, and I will instruct you in what to say." 13 But Moses again pleaded, "Lord, please! Send anyone else." 14 Then the LORD became angry with Moses. "All right," he said. "What about your brother, Aaron the Levite? I know he speaks well. And look! He is on his way to meet you now. He will be delighted to see you. 15 Talk to him, and put the words in his mouth. I will be with both of you as you speak, and I will instruct you both in what to do. Exodus 4:10-15 NLT

Although Moses tried to hide behind somebody else to cover his insecurity, eventually, Moses could no longer hide behind Aaron. He had to face his own insecurities and do what God called him to do despite them. We live in a technologically advanced society where almost everything has a digital reality, but you can't autocorrect and reset your insecurities. The unfortunate reality is, the only person who can overcome insecurities is the person with the insecurities, not the spouse, not the friends, not the parents, and not the partner. You must address and deal with each of them. You have to stop forcing other people to carry, unpack, and deal with the things that belong to you. The more we can become conscious of our insecurities, the more apt we are to find out what we need to work on.

A married friend and I were talking about insecurities one day when she shared with me that when she was single, she thought being married would solve her problems with loneliness. She said that she thought if she found "the one" and that if she made him happy, he would in turn make her happy. She confessed that she was a very insecure woman growing up and thought if she got married her insecurities would fade away. Marriage was her solution to being secure and fulfilled, she confided in me. But her reality now that she is married conflicted with her aforementioned philosophy. She was still lonely and her insecurities and fears plagued her more than ever.

I was reading a singles devotional that said, "To expect another person to make you feel happy, secure, and fulfilled will leave you disappointed at best and disillusioned at worse. Even a great husband

> NR
> "To expect another person to make you feel happy, secure, and fulfilled will leave you disappointed at best and disillusioned at worse. Even a great husband makes for a poor God."

makes for a poor God." The existence or presence of a relationship will not heal nor cure your insecurities. We should never operate from the expectation that if we find that perfect person he or she will right all your wrongs and erase all your insecurities. The error with this thinking is that it will eventually put pressure on your spouse or significant other. Only God can help us with our insecurities. We need to learn to carry our own burdens to the altar and give them to God, not our partner.

I AM NOT MY INSECURITY

I was researching about insecurities when I came across a captivating article of a man who about 15 years ago decided to abandon his lucrative job and travel the world in order to examine himself and to better understand his insecurities. After a four-year journey, he returned home and asked his friend Amanda to participate in a personal project by displaying her greatest insecurity proudly on her body. She did. She wrote thunder thighs across her hand with a permanent marker.

Only God can help us with our insecurities. We need to learn to carry our own burdens to the altar and give them to God, not our partner.

When he posted his photo of Amanda to his blog he got a resounding response from the thousands of people he met while traveling the world. People began to take photos of themselves displaying their greatest insecurities across their body in permanent marker. They also displayed in captions, "I AM NOT," coupled with whatever their greatest insecurity was, across their bodies.

A young Italian woman wrote the pronunciation of illiterate across her forearm "[ih-lit-er-it]" and in her captions she wrote, "I am not my dyslexia" as to indicate her greatest insecurity was triggered by her reading disorder. A caucasian man wrote across his hand, "Am I of no value?" and his caption wrote, "I am not my success" as to denote his value in life should not be determined by the career path he chose. For some, this wouldn't be an insecurity but for him this amounted to be his greatest insecurity. In another photo, an African-American male wrote across one arm "damaged" and across the other arm "goods." His caption read, "I am not my past." For him, whatever he experienced in his life obviously continued to haunt him to the fact that he no longer wanted to be defined by it - his greatest insecurity. There was an older woman that drew a picture of one female stick person on one hand but on the other hand she drew a picture of what appeared to be a man, a woman, and three children, a family of stick people; her captions read, "I am not my envy." Such a powerful statement for these people and so many more to have the courage to articulate and exhibit their greatest insecurities across their bodies, but to also have the courage to forecast that they were not their greatest insecurity.

After viewing these people and their ability to identify their insecurities, I felt compelled to begin my own project. I decided I would exhibit my greatest insecurity across my body in permanent marker. I did this in conjunction with a sermon series called, "Fatal Distractions." I stood before my community teaching about insecurities when I unveiled large letters plastered across my forearm that read, "I am not my father."

Sometimes what plagues me most is the disturbing thought that I will someday be an absent father and a bad husband. I know, it sounds terrible, but I am being very vulnerable right now. The fear that

Section One: THE NEW NORMAL

I won't be able to protect my family makes me terribly insecure at times. Growing up without a father who practically lived 20 minutes from the town where I was raised has proven to have had its reverberating effects.

With repetitive stints in prison and with several children in other families, he didn't have the time or the desire to engage me. So, it's no wonder when I met my ex-fiancé's father who happened to be the second highest ranking official in the FBI counterterrorism unit in the country and a terribly outstanding father, I was a bit overwhelmed. My insecurities were exposed like fireflies on a dark summer night.

Across the back of my other arm I wrote, "Poverty." Being raised well below the poverty-line as an adolescent, I have spent extended amounts of time living out of motels and I've even been homeless as a teenager because my mother was incarcerated and my family grew distant. There I was standing before my people divulging inner secrets that protrude the facade of my adequacies and a resume reading Duke and Princeton which I have learned to successfully hide behind, and how my insecurities have at times handicapped my social life, and at other times, my romance.

Aside from identifying, articulating, and carrying your own insecurities to Jesus, you must learn that there are some things about you that God, the Potter, specifically designed to be the way you are.

So, how do you overcome your insecurities, you ask?

It's not about overcoming your insecurities as much as it is about changing your perspective of how you perceive what your insecurities are. You can only change your perspective by first learning to accept the flaws in your life that you cannot change as God's design and purpose for you. Aside from identifying, articulating, and carrying your own insecurities to Jesus, you must learn that there are some things about you that God, the Potter, specifically designed to be the way you are.

What do I mean by that?

Despite what most people think about Moses and his greatest accomplishments of leading God's people across the Red Sea and through the wilderness, Moses never stopped stuttering. Moses' tongue was still heavy and he still had a difficult time with his speech. However, Moses learned to embrace God's design for him and he still lived his life doing the best that he could with what God had given him. God didn't go back and change Gideon's past; at least, not to my knowledge. Like myself, Gideon still came from a poor family. He still had a socioeconomic and educational background that he was not proud of. Actually, he was still insecure and unsure of himself. Despite his past never changing, Gideon still led an army and he accomplished a great mission of defeating many armies and leading God's people to freedom. Furthermore, Nehemiah, as some scholars have indicated, was the shortest man in the entire Bible, yet he was not given a magical growth spurt before he could redeem God's mission for his life. Despite his insecurities, he, as well as the others, learned to accept the flaws in their lives that they could not change as God's design and purpose for them.

The same goes for you and I. When I look in the mirror at myself, sometimes I have to quote Psalms 139:14, *"For you created my inmost being; you knit me together in my mother's womb. I praise you because I am fearfully and wonderfully made; your works are wonderful, I know that full well."*

> **NR**
> Despite what most people think about Moses and his greatest accomplishments of leading God's people across the Red Sea and through the wilderness, Moses never stopped stuttering.

Do you know this? That you are fearfully and wonderfully made? The Potter makes no mistakes. You have to believe that you are enough just the way you are. You are beautifully and wonderfully made. You are not defined by your insecurities. In fact, you are worth more than your insecurities.

You can change your perspective about what you perceive to be your insecurities by learning to focus on God more than your gap. I mentioned before that I have spent most of my life feeling like I have been drowning in the gap between who and where I am versus who and where I think I need to be. We give birth to our insecurities when we focus on our gap more than our God. The truth is, we all have gaps. Too tall.

You can change your perspective about what you perceive to be your insecurities by learning to focus on God more than your gap.

Too short. Too overweight. The feeling of not being pretty enough. Loneliness. The list goes on. God is not concerned about your gap; He is concerned about you.

Our enemy, however, desires for us to get lost somewhere in our gap. He says, "If I can just get Issac to focus on his insecurities more than he focuses on his God, then I can effectively distract him from what God is really trying to do in his life. The worst thing you can do is forfeit God's will for your life because you chose to cling to your insecurities. Unfortunate things take place when we learn the bad habit of living in our gaps. When we allow our insecurities to own us and to dictate how we live our lives, not only is it bondage, it is a slow death.

The woman in Luke 7 who lived her life as a prostitute, spent most of her time operating from her gap. Every day she woke up and gave her life over to living up to the expectations of her voids. Her desire to sleep with men came from her insecurity and the need to use her sexuality to garner favor from others. She had spent so much of her time expecting other men to make her happy and fulfilled and to feel secure that she failed to realize even a great man will make for a lousy God. She didn't realize only God can settle those deep hearted needs and that a man can never do this. She continued to hide behind men and she continued to experience disappointment until one day, she snapped.

There are three types of insecurities. Behavioral insecurities are based upon something we have done in our lives. Our past actions and failures can cause us to become insecure. Talent insecurities are based on what we know how to do. Many times we feel insecure because of the gifts and talents we may or may not possess. Personal insecurities often stem from how we look and feel about ourselves. Low self-worth can be considered a personal insecurity.

Although it is unexpected, I like to use this woman as an example because when she grew tired of operating from her gap, she did four things:

1. She found Jesus.
She learned to stop depending on other relationships to solve and to dissolve her insecurities, and to focus on her relationship with Jesus to fill her voids.

2. She knelt behind Jesus.
This woman was defeated from her past but she learned that in hiding behind Jesus she could now be defined by His past. The men identified her by her failures but Jesus identified her by the love she had shown

Him. Jesus became her security.

3. She served him.
She used what she had, an alabaster box filled with expensive perfume, to serve Jesus. Perhaps she didn't have the gift to sing, dance, write, or to play music, but the one gift she did have was the gift of service.

4. She recognized that she was worth more than her gap.
This woman teaches us that we are not our greatest insecurity. As a matter of fact, our security should be found in our relationship with Jesus Christ.

You are enough. You, not anybody else, must be convinced of this truth. No matter what you've done, Jesus can cleanse you. No matter how people try to define you, you are defined by the work of Jesus on Calvary's Cross. You may feel that you have no talents or gifts but one thing is for sure, you can serve God right where you are. We all have been gifted with the ability to serve.

Questions that you might want to consider as you work through insecurity.

1. What is your insecurity? (Can you articulate it/them?)

2. Why is it an insecurity for you? (Where did it come from? Why does it thrive in your life?)

3. How have you tried to overcompensate and hide your insecurities? (What lengths have you gone to hide them?)

4. Who have you blamed or hurt because of your insecurities? (Who has had to suffer because of them?)

5. *What are you going to do differently about these things you consider to be your insecurities?*

CHAPTER 4

THE UNEXPECTED GIFT

WHY YOU SHOULD EMBRACE YOUR SEASON OF SINGLENESS

I was conducting yet another marriage counseling session and preparing to marry two more of my good friends. It seemed just about everybody in my circle of friends had gotten married and I was still sitting in the single bracket.

"What's wrong with me?" I had to ask myself.

I know, I know, I'm the pastor and I'm supposed to wait, "yada yada yada." I know I'm supposed to lead by example, I need to trust in God, "yada yada yada." I am fully aware of the slogans and platitudes that typically come from those who are married or who have already experienced marriage. All I am saying is, the struggle to be single is real. So, in an effort to jumpstart God's will to meet my expectations, I launched an all-out assault on every online dating site hoping that God would meet me half way. I remember attending a church conference in Tennessee and while returning to my hotel I had the privilege of riding on a bus next to a woman who was celebrating her 41st year of marriage. She was sitting in the seat directly in front of me. My attention was divided, I was occupied playing with my smartphone patrolling one of the dating websites and updating my profile photo. I didn't realize that my colleague, who was recently married, and this woman, had dove into a conversation about the beauty of matrimony until I heard the happily married woman shout, "I'm so glad not to have to date." Everything came to a screeching halt like the sound of a record needle pulled violently across an album. That caught my attention. Then the unexpected happened, this woman turned around and with this graceful grin on her face, she asked, "How're you doing?" I paused because I had no earthly idea what she was talking about. So, I quickly scanned through my mental rolodex to think of the possible scenarios that could be taking place.

I guess she knew I had been having a hard time being single and practicing full-time ministry. I didn't know whether to become defensive because she assumed I had a problem being single or whether to be vulnerable and just go with it. So I escaped my usual facade of self-sufficiency, while in front of my colleague I risked vulnerability. "You know, I have my days. Some good, some not so good. My temptation is impatience." I said. She responded, "Continue to be patient and I'm praying that you get the right woman." So I gestured, "God's grace is sufficient." She said, "Keep telling your self this, sir." This is why I try to remind myself, "God's grace is sufficient" even though I don't fully understand God's will.

There is a gift in my singleness although sometimes it's difficult for me to embrace it. Sometimes I envision myself in the presence of Jesus when he was washing the disciples feet in John 13:7. They could not fathom what Jesus was doing. Peter was reluctant to allow Jesus to do what He needed to do and that prompted Jesus to say, "You do not realize now what I am doing, but later you will understand." This is a powerful statement that encourages me to embrace this gift and not to switch the price tags on the things I should value most in my life.

Section One: THE NEW NORMAL

WHO SWITCHED MY PRICE TAGS?

I was shopping in a discount department store the other day looking for a bargain, as I am accustomed to doing, when I noticed an awesome pair of headphones priced at an unbelievably low price of $29.99. I quickly grabbed the item and jolted to the cash register like a running back trying to score a touchdown - this was an epic game-changer for me. When I arrived to the counter to check out the headphones, I was charged $249.99. I immediately corrected the salesclerk because I knew that price was erroneous. I actually put up a fight because that bargain belonged to me. To my discovery and embarrassment, the item had been positioned in the wrong place under the wrong pricing.

I figured they still had to honor the pricing since it wasn't my fault the merchandise and pricing was misconstrued. It was a little more complicated than I thought. The merchandise had been positioned in the wrong place, not by the store, but by a previous customer who failed to return the item to its former and rightful position.

But why?

Maybe they found something else that captivated their attention? Maybe, they were prepared to purchase this item only to realize that they no longer desired it, so they just placed it in the nearest vacant spot? Or maybe their hands were full and they realized that it just costs too much and they didn't have the time or the wherewithal to return it to the place it should be, leaving the next customer, me, to be left in a state of confusion trying to put together the jigsaw puzzle to find the true value of the merchandise.

Has this ever happened to you? I wonder how long these headphones were there on that shelf, displaced? Obviously they were there for an extended period of time because there were more than one pair of headphones in the same space. They were there so long that others thought that's where they belonged.

Why didn't anybody return it to its rightful place? The customers? The employees?

After I paused to think about it, I realized I was also guilty of shopping for things and later leaving them in spaces where they did not belong, leaving "employees" to clean up my mess-up. I never really knew how much trouble it would potentially cause anybody until I experienced it done to me.

I'm really not talking about headphones or bargain prices, but I am referring to how easily the value of things get confused and as a result, price tags get switched in our lives. Things that are of great importance and seasons in our lives that are meant to nourish and bless us get confused and mishandled by other people leaving us ravished and empty. Sometimes we can become so used to the hurt, lies, misuse, and disappointment that we allow ourselves to remain in the position of defeat for extended periods of time. As a result, the people we engage don't value us anymore than we value ourselves. Because our hopes and dreams have been misplaced and our hearts and emotions displaced, we leave the next person seeking to date us and enter into a romantic relationship, wondering how everything got so mixed up. We fail to embrace and maximize our season of being single to become better versions of ourselves and better stewards of God's will for our lives. I've been responsible for confusing the value of my merchandise and switching my price tags, but I've learned that when you know how much you are worth, you will stop giving other people discounts.

> **NR**
> I've learned that when you know how much you are worth, you will stop giving other people discounts.

Other times, we allow ourselves to believe that we need something "more" than what we already have in order to be happy and fulfilled. As a business marketing major in college, I have studied consumer behavior, the decision-making processes of individuals and groups and how emotions affect their buying behavior. Most marketing and advertising organizations try to bombard us with unrealistic story plots, captivating colors, and nostalgic music to get us to buy into a 30 second commercial, which reinforces that happiness only comes if you do "this, and that, and the other thing." It comes down to one thing: if you continue to show people what they think they want, eventually, they will think they need it.

This is why we desire faster cars, we need bigger houses, we want six pack abs and a sculpted body, and we covet romantic love - all these things will help us experience ultimate happiness. Therefore, we spend upwards of $34 billion dollars each year on cosmetics and beauty products and over $11 billion in the self-help industry to help us manufacture the ideal life. We effectively switch the price tags on what's most important in our lives. God's will, our identity, and our fulfillment are all left abandoned for things we think will bring us happiness and satisfaction.

What things in your life have been misplaced? What values need to be returned to its rightful position? Are there things in your life that you've been giving more attention to that should not consume so much of you? Starting with your life, first, you need to rediscover your value system and your identity. This is the only way you can expect for other people to appreciate and value what you value. Re-switch your price tags.

PETER PRINCIPLE IN ROMANCE

Peter Principle in Business
"A person will only be promoted to their level of incompetence in any business or organization." The text chimed through a text message in my phone. A friend was in Washington D.C. at a business management conference when she was introduced to a theory in business called, The Peter Principle. She assumed I would find this intriguing but also find it relevant to romance and relationships. I couldn't get this phrase out of my head, so I began navigating sources to discover more about this principle. I arrived at a quote, *"If man is going to rescue himself from a future intolerable existence, he must first see where his unmindful escalation is leading him,"* says, Dr. Laurence J. Peter, the man after which the Peter Principle is named.

In other words, if a man or woman is going to keep from hitting the same roadblock over and over again in their lives, then he or she must discover the reason they are choosing the path they are taking in the first place. The quote suggests that a man should first examine himself and make sure that he is truly growing and becoming stronger, wiser, and better as he progresses and is elevated through any organization rather than accumulating artifacts, fancier job titles, bigger offices, and assuming more responsibilities that give off the appearance of growth, success, strength, and achievement.

In 1969, Dr. Laurence J. Peter published a book called, *The Peter Principle: Why Things Always Go Wrong*. The Peter Principle was intended to be his humorous assessment of the business world, which suggests employees are rewarded for their high performance with a promotion but if they underperform in their new role, they will rarely lose their job. Instead, companies will train more employees and hire more assistant managers to help — all to avoid facing the ugly truth that the promotion wasn't really the best decision. So you will find that hired workers will conveniently work around the manager in order to hide his incompetence. Dr. Peter, a psychologist, observed that in management, people will seek promotion after promotion until they end up in a position that requires a different competency level and skill set they cannot meet. As a result, they will underperform and remain disengaged from their work because they have in fact hit their ceiling - their incompetence.

> **NR**
> We are taught to speed date, commit quick, and marry soon. As a result, we either divorce often or remain stuck in a marriage for years and hide behind false portraits of happiness — all to avoid facing the ugly truth that the marriage wasn't really the best decision.

So the relevant and helpful finding for me was simple: you can move up in any organization and be successful, but if you don't address and fix your flaws, you will always hit a ceiling. That is why many people move up the ranks quickly, but end up in a management position where their incompetence and flaws can't be hidden anymore. Even if they decide to move laterally in a company or on to another company, they will still hit the same ceiling if they don't choose to address that "thing." Not all promotions are good promotions.

The Peter Principle is more than just a critique on large businesses and organizations. It's a symptom of a culture that overemphasizes titles and undervalues being connected to the work you are best at. It's a symptom of culture that pushes us away from the roles and jobs in which we would thrive for the promise of a better sounding resume. We are taught to believe that higher salaries and fancier titles are always best for us, when in fact, every new promotion that comes your way isn't always the best for you. Working in a capacity and on a job that brings out the best in you and allows you to thrive is always better than any title.

DEFECTIVE RELATIONSHIPS 101

The Peter Principle is an attitude that transcends the business world and bleeds into our lives of dating and romance. It's Defective Relationships 101. We are taught to speed date, commit quick, and marry soon. As a result, we either divorce often or remain stuck in a marriage for years and hide behind false portraits of happiness — all to avoid facing the ugly truth that the marriage wasn't really the best decision. We hide behind relationships hoping to cover our deficiencies. *The Peter Principle in Romance,* according to me, is that your relationship will only rise to the level of deficiency in your life. If you are insecure, it will eventually show. If you have anger issues, they will find you out. If you have unresolved emotional issues, they will make themselves available at the most inopportune time in your relationship.

Many dating relationships don't turn out to be the experience they hoped for. Newly cemented couples and even newlyweds find themselves frustrated and hopeless because their expectations and needs are not being met. We enter into relationships based upon our overbearing desire to have a new "title" or because of the "idea" of happiness but we soon hit the same ceiling we have always hit in all of our other relationships because of our unresolved issues, or deficiencies. My transliteration of this principle in romance is you can transition from relationship to relationship and even get a promotion in your status from single to married, but eventually you will have to deal with your deficiencies. If you don't take the time to deal with your issues, no matter where you go and who you date you will always hit a ceiling.

Also, because we overemphasize the need to be in romantic relationships, we undervalue the gift of being single. You will always get stuck underperforming in relationships with disengaged partners if you refuse to take advantage of the gift of being single. So many people are ready to move to the "next big thing" when the next big thing should be your season of being single. Many times, we jump to the next job, the next place of residence, and even the next romance without taking the time to develop right where we are and becoming the best possible version of ourselves we can become. For this reason, we can move 20 times but will always reach a ceiling because our flaws will be revealed and will prevent us from moving forward. Your season is a beautiful gift.

In my devotion, I came across a narrative in I Kings 1:5-9 NLT that gave another example of how our overbearing desire to be promoted and to move to our next season can handicap and thwart God's plan for our lives, even in romance.

5 About that time David's son Adonijah, whose mother was Haggith, began boasting, 'I will make myself king.' So he provided himself with chariots and charioteers and recruited fifty men to run in front of him.

6 His father, David, spoiled him, never once reprimanding him. Besides that, he was very good looking and the next in line after Absalom [to become king]. 7

Adonijah took Joab son of Zeruiah and Abiathar the priest into his confidence and they agreed to help him become king. 9 Next, Adonijah held a coronation feast, sacrificing sheep, cattle, and grainfed heifers at the stone of Zoheleth near the Regel Spring. He invited all his brothers, the king's sons, and everyone in Judah who had position and influence."

Adonijah, the eldest living son of David was extremely eager to take over his father's throne and to enter into a committed relationship to become king over God's children, although he was not qualified and ready be a king. In his impatience and pride, Adonijah did two things: he decided he was going to make himself king so he found some people to corroborate his desires. He also decided to hold a huge coronation feast and to invite all the town's people so they could witness his ceremony in becoming king.

Although Adonijah later failed to become king, his eagerness and overbearing desire to obtain a "title" and to enter into a "committed relationship" exhibits the very nature of the Peter Principle that often pervades our romance. Whether the committed relationship is being a king, becoming a husband or wife, or even engaging in an exclusive romantic relationship with another person, the principle remains clear: you can't customize God's will to fit your desires. Adonijah knew it was God's will for Solomon to become the next king but in his selfishness he wanted to make himself the new king. Another principle remains clear for Adonijah: a coronation won't hide the deterioration.

Adonijah put all his effort in trying to convince people to make him a king. He spent money, time, and energy thinking that becoming a king would be the answer to all of his problems, but before you promote yourself king or queen over somebody else's life, be sure to fix your flaws. Deal with your deficiencies, first. Rushing to marriage won't secure your future. Getting married will not cover or hide your insecurities or your frailties. Too often we attempt to use relationships and marriages as crutches to walk and barriers to hide behind. Adonijah was insecure and very jealous and after time, everybody discovered it. His attempt to become king didn't work out.

The Peter Principle in Romance teaches us:

1. Not all promotions translate into great experiences.

In other words, a nice boyfriend doesn't automatically translate into a great husband. You must operate with more caution and understand that just because somebody performs well in one area doesn't mean they will excel in marriage. Marriage requires a different level of competency, intentionality and devotion. Not all good people make good relationship partners. —— Instead of filling the position with people who seem that they will be a good boyfriend, you should wait and allow God to bring the best person to fill that role.

> A nice boyfriend doesn't automatically translate into a great husband.

2. Do NOT exercise the Power Of Pull.

In his book, Peter suggests that if a manager decides to take the fast track up the ladder and realizes he or she has reached their level of incompetence to attach yourself to superiors who can help pull you up quickly. It's almost a tactic of manipulating your way into a promotion. I want you to do the exact opposite. In relationships we tend to hide behind our partners hoping they will hide our flaws and that we can hide our issues long enough to "get a promotion." If you fail to deal with your flaws they will eventually be revealed. Don't exercise the power of the pull but practice the power of genuinely becoming better.

3. You are always better off in a relationship that brings out the best in you and that enables you to thrive.

No matter how long it may take. It doesn't matter how long you have been waiting. You are always better off in a place where you honestly can thrive and grow. Don't settle for a marriage just because you don't want to be single any longer. You will have a bigger house and more money but you will be less happy and certainly less fulfilled.

> You are always better off in a relationship that brings out the best in you and that enables you to thrive.

CHOOSING ME BEFORE WE
I Corinthians 7: 17 (MSG)

"17 And don't be wishing you were someplace else or with someone else. Where you are right now is God's place for you. Live and obey and love and believe right there. God, not your marital status, defines your life. Don't think I'm being harder on you than on the others. I give this same counsel in all the churches."

A Letter To Myself

Dear You,

Although it may seem contrary, where you are at this moment is a gift. Embrace it. Stop trying to convince yourself that you need to be somewhere else or that you need to be somebody you are not. You don't need the approval of any human being to have value, your worth begins with you. You are the starting point. Stop trying to live the life that other people have planned for you; stop trying to live up to the expectations of others because you will fail every time. You've probably made a mistake or two, or three, but everybody has and will make mistakes, so don't allow your lapse in judgment to define you. Pick yourself up, now. Do me a favor and choose you.

> **NR**
> Call your fears out from hiding. Reveal them and make them known. Specify what they are and inform them that they no longer have power over your life - in the name of Jesus.

What do you have to lose by choosing to love yourself and to focus on you? What's the worst that can happen? You become better? Stronger? Wiser? Whole?

Look beyond the debt, the heartbreaks, the betrayal, the addicted mother, the absent father, the detoured hopes and delayed dreams. Focus on the fact that you have another day and another opportunity to redeem the time. Stop worshipping and grieving your mistakes and keep walking forward. Call your fears out from hiding. Reveal them and make them known. Specify what they are and inform them that they no longer have power over your life - in the name of Jesus. The fear of rejection. The fear of being alone. The fear of trusting yourself again. The fear of failure. They don't belong to you anymore. You will no longer be needing their services. They have been evicted and no longer have your address, so they must be forwarded to Calvary's Cross - let Jesus handle it.

Remember, sometimes you have to take your hands off of certain people and certain relationships and surrender them to God. I know it's difficult to let go but in order for God to do what God needs to do, you must relinquish your control. You cannot walk through the doors that God is trying to open for you until you walk out of the doors that God is trying to close in your life. Doors left cracked are hearts and lives left open, vulnerable, and available for ruin and could potentially ruin you. You have to remember, it's hard to walk in purpose when you have not cut the umbilical cord to the things and people in your life that are nourishing your handicap, weaknesses, and fears.

> **NR**
> It's hard to walk in purpose when you have not cut the umbilical cord to the things and people in your life that are nourishing your handicap, weaknesses, and fears.

Wilderness experiences always demand the person or people in the wilderness to trust God. They force you to make decisions. I know it is counter-cultural, but in choosing yourself, you are indeed choosing the wilderness. You need the wilderness. Don't reject your time in the wilderness. It's when you're in the wilderness that you discover who you are and what your life is really about. You may not have the comfort of a crowd in the wilderness but you will always have the presence of God with you. In the end, God is who and what you need. Practice focusing on God and not anybody else.

> **NR**
> Any person who wants to lead an orchestra must first be willing to turn their back on the crowd.

You have to be strong enough to choose you. You have to be dedicated to yourself. Stop cheating on yourself because you're focused on other people. Any person who wants to lead an orchestra must first be willing to turn their back on the crowd. You have to make this your season. It's about you. Most of all, during this period, you need to focus on picking up the pieces of your own life. The temptation to medicate yourself with people and things to cover up the pain rather than deal with it will be ever-present. Clean out your closet and forgive yourself of the mess you made. Stop adjusting your eyes and your thoughts to the level of deficiency and deficient people around you. When people say, "You have changed," what they are really saying is that you are no longer living the way they want you to live. Read between the lines and take satisfaction in knowing this. Choose yourself. You deserve your own undivided attention. God is trying to do something in your life and God needs you fully available and present in order to do it.

I have to leave now and go fight for me.

Sincerely,

You.

I CHOOSE ME

"So Peter got out of the boat and started walking on the water and came toward Jesus."
Matthew 14:29 NRSV

In this Bible story, Jesus' disciples find themselves caught in the middle of a storm while they're in the middle of the Sea of Galilee. It is dark so the disciples don't notice Jesus when he comes walking on the water toward their boat. The disciples arrive at a crossroads, they have to risk remaining in the boat or they have to risk abandoning their ship - neither decision is easy to make. Peter chooses to leave the boat and to walk toward Jesus. Ultimately, he gets distracted and sinks but he chooses to walk toward Jesus, nonetheless.

I like Peter although I think he gets a bad rap. His life was plagued by faux pas and hasty decisions, but there is something about his life that we can learn and apply to our own. Contrary to popular opinion, Matthew 14 is not about Peter's folly, arrogance, or his inability to trust Jesus. Matthew 14 teaches us how we can choose ourselves instead of choosing others. In choosing to walk toward Jesus, Peter was placing the highest premium over his own life.

You do the math. There are 12 people in the boat but only 1 gets out and attempts to walk toward Jesus. How many do you have left in the boat? Why? Why don't the other disciples walk toward Jesus when Jesus tells them to "Come?"

Peter teaches us:

1. You can have so much going on in your life and be surrounded by many great people and accomplishing great things but still not have Jesus inside your boat. Therefore, when you take inventory over your life, if you are missing the most important crew member, Jesus, either you need to jump ship or you need to prepare to throw some people overboard.

2. If God is going to perform miracles and do something great in and through your life, you must first be willing to disengage from certain things and certain people and get out of the boat of your own comfort zone.

3. Sometimes the people closest to you can serve as your greatest distractions. You need to be careful who you have in your boat because God needs your undivided attention.

Peter arrived at a juncture in his life where he needed to abandon ship or cast some people overboard. In Peter's case, he chooses the former. At the end of the day, Peter accomplishes something that nobody else in human history was able to do, besides Jesus; Peter walked on water. God was able to do a miraculous thing in Peter's life because Peter, in essence, put himself before other people. What does it mean to choose me before we? It means giving yourself the space to focus on God by freeing yourself from fatal distractions. It means giving yourself the space to be Restored, Renewed, and to Reset before you consider engaging in a relationship or an emotionally attached friendship with someone else. Because healthy relationships require emotionally, socially, and spiritually fit people, it's your responsibility to focus on yourself instead of focusing on somebody else.

Most people will say, "I'm not focusing on being in a relationship with anybody anytime soon" but then one morning, they awake only to find that they have accidentally become emotionally attached to somebody. Although it may never be our intention, we must remember that our priorities are not proven by our intentions but by our decisions. The result of choosing other people before ourselves is a lack of self-awareness, purpose, and direction. When I conduct singles' conferences and seminars, I typically ask one or two questions which are almost always met with annoyance and frustration: what is God calling you to do? What is your life's purpose? The purpose of this question is not to make anybody feel insecure about themselves, but to show how we can fail to know our own purpose and yet we seek to find purpose with and in other people. It is counterproductive to attempt to create purpose with somebody else if you don't already have purpose yourself.

If you don't get this, it doesn't matter who you date or become one with because it will only be a fraction of the relationship that God intended for you to have.

CLOSING THOUGHT

"One reason so few of us achieve what we truly want is that we never direct our focus; we never concentrate our power. Most people dabble their way through life, never deciding to master anything in particular."
-Tony Robbins

Pray this prayer:

God, order my steps. Filter my thoughts. Create in me a new attitude. If my circumstances never change, help me to change my perspective about my circumstances. Help me to be a good steward over my season of singleness. Teach me to be content but not complacent. Grant me the wisdom to discern the difference is in my attitude and my expectations. Help me to discover my gifts and talents and teach me how to imitate a healthy tree by bearing good fruit with my life

Amen.

Section 2
RECYCLING BAD HABITS

"No problem can be solved by the same level of consciousness that created it"
Albert Einstein

"No toxic relationship can be cured by the same level of deficiency that created it."
Issac Curry

CHAPTER 5

YOU DON'T "NEED" TO DATE

DATING AND THE CRISIS OF CODEPENDENCY IN RELATIONSHIPS

Many people overstay in relationships that are subpar and toxic because they are afraid of being alone or they feel solely responsible for the other person's happiness. They may talk about leaving but they never follow through with their convictions. They either stay, return to that relationship, or enter into another relationship that replicates similar emotional experiences. It's almost like an addiction. It's hard to follow through with the breakup. It's even more difficult to be alone and to keep from becoming emotionally attached to anyone for an extended period of time. They are seemingly drawn to that same type of person who produces the same types of behavior as their previous partner. For some people, this conflict is unfortunately home for them. It may bring about a certain level of stress to engage and remain in these types of draining relationships but there is an emotional dynamic within them that is accustomed to dealing with emotionally distant and unavailable people. It's a cycle. This excessive emotional reliance on another person or partner is called, co-dependency. People who exhibit traits of codependency often put a premium and priority on tending to other people's needs rather than their own.

Originally, it was difficult to identify and more embarrassing to confess, but I understand now more than ever that I was predisposed to codependency growing up - it flows through my veins. This is something with which I currently wrestle and consciously work hard to avoid nourishing. This learned behavior, as researchers contend, is something often garnered from childhood and family experiences. According to gifted psychotherapist, James Hollis, I was a "parentified child." In order to overcompensate for the disorders and absence of my parents, I tried to take on the responsibilities of the adult in the family as an adolescent. I became accustomed to enabling the habits and the destructive behaviors of the people closest to me. I thought remaining present was showing my love and support when in fact it was an exhibition of a mutually dependent relationship in which I was learning how to feel wanted in order to avoid dealing with my own insecurities. This is the anatomy of a codependent person. When a codependent person enters a relationship, the relationship often becomes codependent, destined to repeat codependent relationship patterns. Instead of growing together, these relationships often deteriorate together.

Many people in codependent relationships:

1. Have very poor boundaries. One or both persons have a difficult time saying, no, because of the fear of hurting the other person and being disliked.

2. View loyalty as their reason to remain committed to relationships that are abusive and that do not meet their needs.

3. Have a deep-seated need to be needed; they thrive on caregiving and rescuing and teaching and solving the problems of other people.

4. Are dependent on the approval from the other person or people for identity and self worth.

5. Are in the relationship primarily because of their need to have some form of an emotional attachment to something or someone. These people always feel a "need" to be dating someone.

RECYCLING BAD HABITS

The greatest gift you can give somebody is your own personal development. I used to say, "If you take care of me, I will take care of you." Now I say, "I will take care of me for you, if you will take care of you for me." These profound words were shared by author and motivational speaker, Jim Rohn. When I first read these words, I remember having to take my hand and physically close my mouth. It was an unexpected rebuke. It was enlightenment. These words shifted the paradigm from which I operated my entire life. It forced me to rethink everything I thought I knew. There are

The greatest gift you can give somebody is your own personal development. I used to say, "If you take care of me, I will take care of you." Now I say, "I will take care of me for you, if you will take care of you for me."

several habits and behaviors in relationships that most people think are normal that are unhealthy or dysfunctional. For example, I grew up wanting and envisioning a woman that would take care of me so that I could take care of her in return. That's how I thought the story was supposed to read. This is what our secular culture taught me. On the surface this sounds perfect, but underneath the veil, there are codependent traits woven in the fabric of this mindset. As a result of this thinking, I had inadvertently made the person I was dating responsible for my emotional wellness and the degree to which I was pleased and happy with the relationship. There was always an imbalance in my relationships because what I thought was normal was indeed my codependency on autopilot. All codependent relationships and people don't look the same but there are key ingredients that nourish and provide a platform for them to thrive, two of which are the lack of self-awareness and accountability.

One example is Seth and Marianna. Seth and Marianna had both been married for several years. Seth liked to bend wire and create different trinkets and copper animals in his spare time. He would go to the garage and find peace in this hobby. Marianna thought it wasn't the best use of his time and she discouraged this hobby so he stopped bending copper. Seth didn't want Marianna to be displeased with him and he didn't want her to think of him as being wasteful with his time. Seth later went to the attic and dusted off his banjo from childhood and thought he would begin playing his favorite instrument again. Marianna didn't like the sound of the banjo and didn't like the fact that Seth was spending time away from her, so Seth forfeited this hobby also. In turn she wanted him to spend more time with her watching evening game shows that she liked to watch, so he did this because he knew it would make her happy and she wouldn't be upset and feel alone. He later resented her because all he did every day was sit on the couch and watch game shows with her because he didn't want to appear unwilling and inflexible..

Seth and Marianna exhibit a common codependent dynamic in relationships. One person denies a portion of his or herself just to make the relationship work while the other person has a compelling and sometimes subtle need to control what the other does and how the other person feels. Codependency requires that you deny part of who you are just to make the other person happy or to seek their recognition and approval. There is a subtle way that we give away our power and enable codependency to thrive in our lives. By refusing to say, "No" Seth sought Marianna's approval and didn't want her to feel like he was abandoning her. In turn, Marianna wanted to monopolize Seth's time and she preferred that he spent his time doing only what she liked to do. This couple failed to identify and accept their behaviors as abnormal and they were unable to embrace their differences, so they parted ways.

Three Codependent Behaviors you think are normal but are not:

1. Excessively compliant in the name of being "submissive"
Many times a person who is codependent becomes excessively compliant in a relationship because of the fear of rocking the boat or upsetting the waters. Not only have I personally experienced this, but I have seen it in my household and among peers, primarily women. Here's the catch; many times, a woman who lacks self worth and identity will pull the "submissive" card in a relationship or marriage and use that as a scapegoat to be excessively compliant. Being submissive does not give a person a green light to become so compliant in a relationship that you deny your own voice, happiness, and health. Yielding to your partner to the point that you lose touch with yourself, your desires, and

your needs does not make a relationship healthy or successful but it gives it the fuel it needs to fail.

2. Denying your own needs in the name of "compromising"

We are taught that in order for a relationship to be successful, two individuals must be willing to make the necessary sacrifices for the relationship to work. This is true. However, compromising does not and should not be a one-way street. One person should not consistently deny his or her own needs for the sake of the relationship alone. This is unhealthy behavior that we like to think is healthy. We attempt to justify our need to be needed with our unhealthy willingness to focus on the other person's feelings and needs without ever taking into consideration our own needs. People who are codependent or enable codependency call this compromising. Healthy relationships require mutual compromise in which both parties equally participate. This is called, interdependency. Interdependency is about fostering a climate in relationships that is mutually satisfying and where both partners rely on each other equally.

> **NR**
> Interdependency is about fostering a climate in relationships that is mutually satisfying and where both partners rely on each other equally.

3. Fear of abandonment and separation in the name of "loyalty" and "working hard" to make your relationship work

"You don't care about this relationship." "Why are you giving up so easily?" These catch-phrases will almost always make it difficult to walk away from a codependent relationship. One person feels that the failure of the relationship will replicate her experiences of abandonment and rejection so she is practically willing to tolerate mayhem for the rest of her life if it means she won't have to be alone. The other person doesn't want the person to feel abandoned or hurt and feels it is his responsibility to make her feel happy and content. He allows guilt to keep him bound to the relationship. This is classic codependent behavior. We remain in relationships longer than we should, we tolerate more than we can bear, and we deny ourselves freedom and happiness because our insecurities have held us hostage.

"NOURISHING MY CODEPENDENCY"

[journal log]

One day I was standing in the mirror on the opposite side of the wall in my dormitory room looking at whether my weightlifting was paying off. From the depths of the mirror, I caught eye contact with the person staring at me. This encounter was unlike any before. I tried to look the other way but he kept whispering my name desperately trying to catch my attention. I looked up but this time I couldn't turn away. I walked a little closer. I noticed tears had begun to well in his eyes. He said, who are you? There was silence. He spoke again, you can stop performing now. Nobody is around. Lower your facade. I tried. You can't carry on like this, he said. I shrugged my shoulders and pasted a smirk on my face politely dismissing this matter; stop talking to yourself, I nervously spoke. However, it didn't work, at least not this time. Like a timed bomb having patiently awaited its moment of truth, he burst into an ugly cry. In desperation, he clung to the nearby wall to help him stand. The look in his eyes as he patrolled me was that of desperation. His eyes remained affixed on me like a bull to a matador. In angst he bellowed, people don't really know you, they don't know what your life has been like. They don't know the price of your ticket. The sleepless nights; the domestic violence; the substance abuse; the parentless home; the homelessness; the poverty; the depression; the fear; the womanizing; the manipulation; the lies; the pain, all invisible to the naked eye. You are like the invisible man, he said. Everybody sees an affable, burgeoning, and well spoken young lad, but underneath the veil is a putrefying stench that only he could smell. Through the orifices of his pores reeked a sensorily unpleasing odor coming from his close acquaintances: denial, self-indulgence, codependency, and sin. Perhaps this explains why he was alone in this room, one thousand miles away from home. God needed to get his attention. God needed to move him out of his comfort zone to begin to separate him from his vices.

The three years I spent in graduate school in Princeton, New Jersey proved to be one of the most eye-opening experiences of my life. I was stretched intellectually. I engaged the minds of world-renowned philosophers and theologians. I was exposed emotionally. Living in New Jersey, was the furthest I had ever ventured from my family and friends. The culture was different. The people were different. Even the weather was different. The school-work was extreme, adjusting to cold weather was insane, but believe it or not, nothing was more perplexing than not being able to foster a romantic relationship with somebody.

Confession. My time at Princeton was the first time in my life that I didn't have somebody readily available to nourish and enable my need for an emotional attachment. I couldn't meander through my phonebook and make a phone call or send a text message. I didn't have the ability to go up the street or around the corner. I was one thousand miles away from my comfort zone. This is when I first discovered I had a problem. I remember sitting cross-legged in the center of my floor in my room and pulling out a piece of paper to write down the names of the women I had dated in my past when I came to a disturbing discovery; there had not been a time in which I was not in some type of relationship with a woman. Get this. It took me to be miles away from home and completely out of my comfort zone to realize I had an unhealthy need to always be in a relationship. At this time, I wasn't fully aware of terms like codependency and how I had been using women and my charisma to nourish my codependency. I was addicted to relationships and like any addict who abruptly discontinues his use of a substance, I was going through withdrawal. I was irritable, angry, and questioning God's providence, if I would ever get married. I was, as my close constituents would put it, thirsty. I was like a stray feline roaming the city gravitating toward every person who remotely showed an interest. I tried to pursue relationships but nothing surfaced. I didn't as much as get someone to agree to have a cup of coffee with me. I was alone and could no longer lie to myself. I was lonely.

I know a woman named Sidney who also had a habit of lying to herself. Like me, she exhibited all the characteristics of an addict. She had spent so much of her time trying to find a man to make her happy and to fill her voids that she completely lost herself. She wanted to be happy so badly and to replicate the happiness she saw in others, she figured she would get married. She wasn't completely in love with her husband. She got married to him because she wanted to prove that she was worthy of marriage. They divorced. Sidney got married a second time. This husband she loved. However, they were worlds apart. He was from Mars and she was from Venus. He wasn't a Christian but he looked good on paper. He was good-looking, had a great personality, and he could provide her with a great lifestyle. They later divorced. They had two different value systems and truthfully he didn't really want to be married anymore.

Sidney met her soon-to-be-third-husband. He told her he loved her and that she was the most beautiful woman he had ever met. He expressed such an interest in her, that she was sure he would make her happy. She cooked. She cleaned. She worked two jobs to support the home. Over the course of eight years, he became emotionally unavailable and abusive but she was committed to stay because he loved her and they were together for such a long time. Eventually, the marriage ended. She married him out of convenience and because she liked the fact that he wanted to be with her. Sidney's fourth marriage was over before it began. There was no ceremony. It was more of an agreement. There was no real intimacy although there was a lot of sex. She was only attracted to him physically.

After her fourth marriage, she gave up on being married and decided to just have a live-in boyfriend. "Why go through the stress of marriage again," she concluded. She was fully satisfied with her current boyfriend; she was still thirsty and seeking a man to fulfill her needs. One evening, she left home and told her boyfriend she was going to the store but really she was leaving the house to seek to find the next candidate to play the lead character in her life. This night, she found the right man, Jesus. In John 4, Sidney (the Samaritan woman) ultimately finds Jesus and from this point, her life is never the same.

In John 4, the character had been married and divorced four times. The man she was currently living with wasn't her husband. Sidney was found trying to fill her water pot in a location that was an unusual distance away from her home. Jacob's well was a very popular place to connect and to meet new people. Some believe that she traveled to great lengths because she was hoping to meet and find a connection with a man. Her plans didn't quite work out that way. She didn't find what she wanted, she found who she needed - Jesus. Like this woman, perhaps you're exhibiting all the tendencies of an addict. You constantly feed yourself lies. The shoe doesn't fit yet you're still trying to force it. It's an unhealthy relationship but you lie to yourself because you're addicted to filling your voids. You have a gap that he's filling in your life and you're hooked. You can't handle the idea of separation. You seek that emotional connection because you feel you need it. The late Myles Munroe once said, "If you feel that you need to date, that's probably an indicator that you are not ready to date and that you need to wait." Wherever there is a "need," this implies there is a deficit and deficiency somewhere. When we seek relationships and companionship because we feel we need it, we run the risk of building that relationship around our deficit and our need. As a result, the relationship becomes imbalanced. When we seek to find other people to provide us with purpose and happiness, we set ourselves up to lose.

When we seek to find other people to provide us with purpose and happiness, we set ourselves up to lose.

When you leave the care of your identity in somebody else's hands and when you base your self-worth on what people think about you, you've become an indentured servant. Why sell yourself into bondage? Stop nourishing your need for a relationship and begin fostering independence.

CASTING CALL

She posted on my blog for all to see, *"So I have had to face that men are an idol in my life and I have had to die to putting time and energy into online dating. It is so hard to control my thoughts and die to self. I have always had a man in my life. This is so hard. I'm in a battle for my life. I've been facing this and failing for over a year now, and of course none of the relationships ever work out because they were never meant to be. I need prayer. God has to be first...and I need healing. I will do okay for a while and then I cave in and just want that attention from a man. Can you please pray for me?"*

I couldn't resist. Her words made me stop everything I was doing at the time I came across this posting. Her humility and the freedom to speak freely made me respond to her, openly. "You, my friend, are very courageous to reveal the inner struggles most of us currently have. I appreciate you more than you know for risking to be vulnerable. Your words are heartfelt and authentic. We don't glory in our struggles but it blesses others when we are able to be open and real about our lives and our frailties. I, too, join you in this struggle. Make room for me at the altar. Jesus, have mercy." She later explained to me that ever since the age of 14 she had never been alone and would have another boyfriend lined up as soon as the other one ended. It was after the death of her husband, Todd, that God had begun to reveal to her that men were an idol to her and that He desired to be the One and only in her life.

Just like the woman at the well and so many others, we waste entirely too much time trying to play the casting director in a movie that is not ours to begin with. We desperately seek to hire performers to entertain our flaws, romanticize our fears, and to tell us what we want to hear so that our movie script reads "happily ever after." And while we are so busy trying to manufacture a happy ending with somebody, we end up getting stuck with a character who should have been written out of our scripts a long time ago. "Casting call! Anybody want to memorize the scripts of my needs, fears, and my codependency?" This is our mantra. Instead of relying on the direction of God, we decide to play the role of central casting and we permit the wrong people to have access to our backstage, we even allow them occasionally to sit in the director's chair. As a result, they control our emotions - whether or not we will have a good day or a horrible one. They say, "Action!" and we smile and perform and act like we are happy but in all honesty we just don't have the courage to walk away. They say, "Cut!" and we alter our goals and our desires to satisfy their expectations because we want to make them "happy." Cancel the auditions. Better yet, you may be preoccupied with trying to become the leading character in somebody else's drama. The sad part is, when we can't get the lead role, many times we settle for the role of an extra, body double, or stand-in, who has no contract, no assurance, only hopes and aspirations. "I will perform for affection" is what your actions communicate when you disregard the voice of God, lower your standards, refuse to say, no, to the wrong people and yes to the will of God. You have to cancel the auditions.

JESUS DEFICIENCY

When I was young, like most children, I loved to stay outdoors. If I had permission, I would leave in the morning and my mother would not hear from me until the streetlights came on indicating it was time for me to come home. On many occasions, I would go outside and because I was so focused on playing with my friends, I would neglect essential things like food and water. I remember one Saturday morning I left to go outside to play. I returned home, midday, because I had fallen sick. Like most boys and men, when we get sick, I was very dramatic. I may have asked my mother to take me to the hospital because it was an emergency. I explained to her that my stomach was hurting, I had a headache, and that I was severely dizzy. I asked for Pepto Bismol or something to quiet my stomach. I pleaded for some type of headache medicine and something that would help my dizzy spell. I needed anything that would make me feel better. After a short period of fighting with my mother trying to diagnose my illness, she asked me a question very calmly, "Are you hungry? Have you eaten?" I replied, "No, I can't eat! I'm not hungry!" I continued my rant about how sick I was and how I needed medicine. She left the room and later returned with a tuna sandwich and some potato chips crunched on the inside, the way I love it. She told me to eat the sandwich while she prepared the medicine. I did. When I finished eating the sandwich, you wouldn't believe it, suddenly, all the symptoms left me. It was a miracle. About the time my mother returned, I was outside playing football again. She never prepared any medicine for me. She knew my problem wasn't that my stomach was hurting because that was merely a symptom. She knew my problem wasn't that I had a headache because that was a symptom, too. My mother was keenly aware that the real problem was that I lacked substance. When I received substance, every symptom subsided.

> **NR**
> If you never get married, if you never get that love that you have always longed for, if you never get healed, if you never get the answers to your prayers, if your circumstances never change, is Jesus Christ enough?

The problem with many of us is that we are "outside" and because we are so focused on playing with our friends, earning our degrees, building our lives and looking for love, we neglect the essential things that we need for survival. We meander around with our aches and pains and hurts and heartaches trying to cure our symptoms, thinking our problems are in one area when in fact, we are hungry for the substance of Jesus. There is a Jesus deficiency and God has been trying to get your undivided attention so that He can provide you with what you really need. We have to stop trying to cure our symptoms and address the real issue. Our problem isn't loneliness, that's merely a symptom. Our problem isn't low self-esteem and codependency, those are only symptoms. When you incorrectly diagnose your problems, you will always prescribe the wrong medicine or take the wrong approach. When we are deficient of Jesus, the only person or thing that can help us is Jesus.

Ask yourself this question. If you never get married, if you never get that love that you have always longed for, if you never get healed, if you never get the answers to your prayers, if your circumstances never change, is Jesus Christ enough? Will Christ be enough for you? My godmother, may she rest in peace, always longed to be remarried. For 20 years, she remained celibate. She prayed and mourned and sought the Face of God because she desired to have somebody to love her and with which to build a family. As she grew older and eventually expired due to an aneurysm, I remembered that she used to repeat to herself, "Jesus has to be enough for me."

CHAPTER 6

I LOVED IT WHEN SHE CALLED ME "DADDY"

DATING AND THE TANGLED WEB OF TRANSFERENCE IN OUR RELATIONSHIPS

WHAT IS TRANSFERENCE?

I don't profess to be a psychotherapist. That's why I have sought the guidance of several specialists in order to briefly discuss this matter. David Richo, in his book, "When the Past is Present," offers advice on the subject of transference. He defines transference as, "An unconscious displacement of feelings, needs, attitudes, expectations, biases, fantasies, perceptions, reactions, beliefs, and judgments that were appropriate to former figures in our lives, mostly parents, onto people in the present." In its simplest form, transference is when we "transfer" or "carry" the past into the present, mostly in our everyday relationships. It is believed that all people will have suffered from some form of transference in their lives. One of the most compelling statements Richo makes is:

"Though most of us want to move on from the past, we tend to go through our lives simply casting new people and the roles of key people, such as our parents or any significant person with whom there is still unfinished business. In transference, feelings and beliefs from the past reemerge in our present relationships. Transference is

Transference is unconscious; we do not realize we are essentially involved in the case of mistaken identity, mistaking someone in the present for someone from the past.

unconscious; we do not realize we are essentially involved in the case of mistaken identity, mistaking someone in the present for someone from the past. Transference is a crude way of seeing what is invisible, the untold drama inside of us…Noticing our transferences may not be so difficult, since we choose people on whom to transfer that really do resemble our parents or other significant characters in our life story. Indeed we can piece together our childhood history from the crypt of our unconscious. We do this by observing our needs and expectations in relationships and the partners we choose or keep choosing… Transference smuggles the past on board the present, and mindfulness escorts us safely to the port of the present, our illicit and burdensome cargo now casted overboard."

Because I was raised in a lower-class, single-parent home with an alcoholic mother who at periods was absent herself, I was forced to wear multiple hats as an adolescent. To my older sister, I was her older brother and mentor. To my younger sister, I was her father, brother, and protector. To my mother, as dysfunctional as it may sound, I was her husband, her confidant, her protector, and at times, her leader. Whenever there was a problem, I had to figure out how to solve it. Wherever there was a need or a proverbial fire, I had to figure out how to extinguish it.

And so it was, I had become a firefighter and a rescuer in my family and it was a badge of honor for me. I was the fixer, and when I fixed things, I proudly stuck my chest out. I had learned early how to be the leader of the household. I never had the privilege of being only a son or just a younger brother. Unfortunately, there was a price to pay for wearing multiple hats in my adolescence that I would consequently have to pay for when I became an adult. I brought that dysfunction from my childhood into my present, and slowly it began to destroy me. I transferred the role of being a firefighter and rescuer in my home and early life into my romantic relationships as I became older. I found, just like in my childhood, I was guilty of wearing multiple hats in my relationships. I didn't realize that I had built an imaginary bridge from my past that led directly into my present. Everyday, I would unconsciously walk to and fro, coping with my yesterday by how I handled my today. I'm guilty of casting roles, responsibilities, and experiences with the women in my family onto the women I was electing to date.

I have always been the man that all the women in my family depended on. I have been the together one, the one with the drive, the one with the plan, the one that makes everybody feel loved. I made everyone feel important and I affirmed the insecurities of others. I covered up their weaknesses. Those became the relationships that manifested themselves in my love life and types of relationships I gravitated towards. These were also the same relationships that brought me the most pain, anger, and fatigue. Although these were the types of women I had always felt an inexplicable attraction for, they were being transferred from my past into my present. As a result, those relationships ended before they even began. This is real revelation for me.

I silently waged war with myself because I never knew why I was so discontent in these types of relationships. I desperately wanted to learn how to be with someone who just wanted to be with me, not somebody who needed me because of a deficiency in their own lives. I did not want someone who was with me because they needed to overcompensate for something they didn't have growing up or in their past relationships. Everybody is guilty of transference but not everybody is willing to identify what, who, and why they transfer what they transfer. This is why relationships suffer.

I'M THE DADDY

November 6, 2009 at 9:36pm, she defined me as a rescuer and she told me that she admired that about me.

November 7, 2009 at 5:01am, while sharing a room, she walked into a dark restroom and flicked the light switch and discovered we did not have any electricity. She thought I was asleep but I wasn't. She whispered, "Daddy, we have no lights."

The first day she called me "daddy" I just knew I was in the perfect situation. I mean, what man wouldn't secretly want his woman to refer to him as "daddy"? We were still in Africa where we had just met. She had been displaced by her missions organization and, by happenstance, I was there to help aid her through the duration of her trip. We traveled the country doing missions and exploring the land together. I loved the fact that I was "daddy" to her. It made me feel strong and useful. I had purpose. I was needed. I guess I really liked the fact that I was needed.

After returning to the States, we continued to spend time and eventually decided to pursue romance. It was like a love story. I was sold. Eleven months later, I proposed to her. During our engagement, I recall occasionally confusing my role as a rescuer and father. Not that she was unable to take care of herself or needed me to teach her anything, that wasn't necessarily the case. She was very, very independent. It's just that the lines blurred from time to time.

After our engagement failed to launch, I departed that relationship pointing my finger at my ex-fiancé. I felt she placed unrealistic expectations upon me. I felt I couldn't possibly live up to the man's man her father was. I honestly felt she had unresolved issues with her father as I recalled private conversations we shared. I felt, she wanted me to be who he was. I tried, but failed. Boy, how my prism was one-sided and jaded. I eventually learned, a couple of years removed from our relationship, that this wasn't necessarily the case.

I was the one that was suffering from a mistaken identity. I presented myself as the rescuer and the firefighter. I wanted to be the one who extinguished the fires and conducted rescue missions in my relationship. I thought that's who she wanted me to be, but I eventually grew fatigued and worn. The problem was, I tired my own self out and wanted to stop wearing the hat I created for myself. I wanted to just be me, the boyfriend. I was emotionally immature and I was guilty of transferring. I unconsciously built that imaginary bridge from my past and I allowed her to accompany me. She accompanied me on this journey that would position her in the role of the other women from my childhood. I was the rescuer. I was the daddy. I wore multiple hats in my relationship until I got tired of wearing them and just wanted to wear one. But by then it was too late. I made my bed so I had to lie in it. Inevitably, I destroyed the one relationship I wanted to cherish. What began as a term of endearment turned out to be an explosive device that detonated when we least expected it.

I was a firefighter and I cultivated the climate for destructive behaviors to flourish in my relationship. I spent so much time putting out fires and being a rescuer that I never really was her boyfriend. The dysfunction of wanting to be a hero handicapped my relationship.

BAD HABITS DIE HARD

I've spent years bargaining with reality concerning my need to be the fixer, the one who saves the day, the problem solver, and the miracle worker and savior of some sort in my relationships. I tried to address it and I thought I had dealt with it. Have you ever dealt with something only to realize that it is ever-present, kind of like an eternal game of Whac-A-Mole? My need to be the hero runs deep into my veins. It is far more potent than I ever imagined. I've come to the realization that I have a deep-seated consciousness that naturally builds a bridge from my past into my present. I unconsciously gravitate towards and attract women that in some way or another are attracted to or need some type of rescuing. There is some level of codependence that enables the relationship to exist. I become so accustomed to wearing the hat of a firefighter and rescuer that I never get to really wear the hat of boyfriend. This is my bad habit. It must die, hard.

In my relationships I have realized that I often assume a false sense of responsibility. Because I like to be needed, I make myself responsible for things and circumstances in the relationship that I am not supposed to be responsible for. Since we are in a relationship together, I typically feel responsible for anything that takes place in the relationship that may keep you from being happy and satisfied. I make it my responsibility for you to experience happiness, when in reality, that is out of my control. I'm starting to realize that I cannot continue to take responsibility for things and circumstances that are not in my power to change. I also realize that I have the tendency to borrow other people's burdens. I force women not to have to handle their own issues, problems, or shortcomings by immediately taking their burdens and making them my own. If you have misery, pain, or issues that need to be fixed, I immediately put my hands to the plow and find a solution for you. I will confront your problems *for* you.

Over time, God is convincing me that it is not my responsibility to necessarily bear other people's burdens in relationships. Although Galatians 6:2-5 teaches us that we should *"carry each other's burdens,"* it also teaches us that, *"for each one should carry their own burden."* It is our responsibility as Christians to help people carry the weight of difficult times when they come, but it is also our responsibility to allow people to carry the cargo they may have on their own ship. It's okay to walk alongside someone on their journey of trying to solve their own personal issues, but it's not okay to attempt to walk out their journey for them.

Let go of other people's addictions and issues and take responsibility for your own life. This way, you live and love as a free person.

God is helping me to identify and bring these things into my consciousness so that these bad habits can be put to rest once and for all. You have to ask yourself, "Is this really my problem to solve? Am I really helping them or encouraging them to become dependent upon me? Is it my place to make the change for them?"

You have to stop borrowing other people's burdens and eliminate your false sense of responsibility. You can't continue to bargain with reality. You can't be somebody you are not meant to be, nor can you make somebody else occupy a role in your relationship they were never meant to occupy. You have to give yourself permission to be you and not somebody else in your relationship and not feel guilty about it. You also must realize that you can love a person without saving them. Let go of other people's addictions and issues and take responsibility for your own life. This way, you live and love as a free person.

We have to stop letting what we are not getting from other relationships in our lives or from our past relationships to control the life of our current relationships. It's the destructive system of transferring that we must bring to our awareness and walk away from. When you rescue others, you usurp their power of responsibility and self-development and cause them to create an unhealthy dependence on you.

"We do not eliminate transference; we decant it. We do not kill it as David killed Goliath. We wrestle with it respectfully as did Jacob with the angel, until it yields it's blessing. The blessing is the revelation of what we missed or lost and the grace to grieve it rather than transfer it. We feel they momentum to mourn all those who did not make time for us, to let go of their importance to us, to go on with life no longer determined or unduly influenced by what others choose to do. We find satisfying sources of need – fulfillment in ourselves and in other humans who can be there for us most of the time and not there sometimes. And in a yes to that, we have all we need." David Richo

THE ELEPHANT IN YOUR RELATIONSHIP

My brother and best friend, Timothy, got married five years ago and I was the best man at his wedding. I remember it, just like yesterday; the ceremony was very small and elegant. Five years later to the date, he calls me and in utter disappointment as he shares with me that his marriage is on the verge of collapsing. "I shouldn't have married you, you are not the one for me." These were the words spewed from his wife's mouth like water from a faucet onto him. One day, in the heat of an argument, she just unloaded everything all at once, "I've been unhappy for all 5 years and I've been masking it the entire time," she said to him. Through all of the ruckus and fighting, he had become frustrated trying to discover the core of her discontent. "It seems like I can do nothing right. Everything I do is not good enough for her," he confessed.

As heart wrenching as it was to hear him vent about his conundrum, something happened. She confided in him, "I've never had anybody to be there for me other than my parents." That was it. His wife had unexpressed unmet expectations and hurt from her past, and that hurt had somehow crept into their marriage, unidentified. This is not to say that my brother did not play a role in anything. Sometimes you can be involved in a boxing match in your relationship or marriage and you can have no idea who or what you're really fighting about. She felt she could not depend on him, financially. What could have been a conversation about unmet expectations, became a volcanic eruption when nobody even knew there was hot lava underneath the ground.

Somewhere I read that hurt people hurt other people. The idea behind this notion is that we often enter into relationships and marriages toting around a worn suitcase filled with artifacts of broken dreams and distorted beliefs. We hope to erect or resurrect them in our current relationships. Like a child building a temporary structure with Legos, we try to build relationships that easily fall apart. We are unwilling to excavate the landfill of manure in our own backyard or dig below the surface of our awareness in order to redeem the emotional safety deposit box which holds the truth to our dissatisfaction and deficiency. We are still holding on to hurt from our past relationships and even childhood.

Hurt people hurt other people because we choose to ignore

the depths of our pain, deficiency, and the depths of our heartache, all while pursuing relationships with other people. Ultimately, we end up hurting them, too. So now the unresolved emotional issues, or elephants, that you refuse to address and act like you don't see, now become the elephants in your relationship that hold you back. They also hold your partner hostage. There is an elephant in your relationship because you brought the elephant along with you and now you are engaging somebody else in a relationship. There are now elephants everywhere that nobody really wants to deal with, though everybody is suffering.

If you pay close attention, the apparent elephant showed its face in their debacle when she said, "I've never been able to depend on anybody but my parents." Somehow their current problems in their relationship got rewired from her past relationships and replayed in their current one. Financial distrust from her first marriage impacted her ability to fully trust her husband in her current marriage. Because Timothy was unable to provide her the lifestyle her family gave her, she became unwilling to compromise any longer. Despite the two and sometimes three jobs he worked at one time and despite the graduate school degrees he earned, he just wasn't able to secure the greatest of jobs in this rough economy. There was an elephant in their relationship.

> Hurt people hurt other people because we choose to ignore the depths of our pain, deficiency, and the depths of our heartache, all while pursuing relationships with other people.

Elephants can be a lot of different things, but let me give you a few examples.

Section Two: RECYCLING BAD HABITS

1. Trust Issues

When you enter into a relationship and you feel the need to tell the other person, "Hey, I have trust issues." Well, here's the deal; the person you are dating has nothing to do with the baggage of your past. They shouldn't be the honorary recipient of an award they didn't ask for because somebody else misused and abused your trust in the past. Now your boyfriend has to wrestle with your elephants.

2. Approval Addict

When you're dating because you want to be received and affirmed so badly by the other person and you become an approval addict. You sacrifice your values, your happiness, and your voice, just to make the other person happy. As a result you lose yourself, you compromise your identity, and you surrender your emotions, and you wonder why the relationship revolves around him? The fact that the relationship revolves around him isn't his fault, it's yours because you allowed it. You wanted his approval so badly that you denied yourself the things you desired and valued. You're an approval addict because you have some elephants that have been aging from your past that you decided to ignore. Now in your current relationship you both are too busy dancing with each other's elephants that you can't simply love one another. Stop allowing your insecurities and your fear of loneliness to sabotage you and convince you that you need to attach your self worth to another person's opinion and approval of you.

3. Distorted View of Love

When you enter into a relationship, they tend to be limited only to physical touch and sex because somewhere in your life the wires got twisted and switched. You think that sex is a sign that he or she loves you or values you. When you look at your dating relationships, they all have had the habit of becoming purely lust driven and sexually dominated. In your past, or even your childhood, somebody somewhere did something to you or took advantage of you and now your wires have gotten twisted. You move from relationship to relationship wondering why they all end the same. This is because you have a distorted view of what love looks like. You have trained this particular elephant to be quiet and to do tricks. Now you are in this relationship, and for the first 6 months, nobody sees the elephant because you've trained the elephant to do what you want him to do. Eventually the elephant wants to come out. You're now wondering why your relationship is going downhill when all you do is have sex to resolve your issues.

4. Instant Anger

Do you instantly get angry when you feel ignored? When you don't get the attention you desire? When you don't get your way? Ever seen that before? You wonder why when you make the slightest remark, he just blows up? Carrying elephants from your past will do this to you. You get tired of carrying them. Training them. Hiding them. Trying to make sense of them. I remember a friend used to describe me as a "quiet boil." Nobody should have to operate at a quiet boil. Instant anger has always been an elephantine issue that I've had to deal with in my life. This is an elephant that you must refuse to tame. You must get rid of it.

5. Mother/Father

Have you ever been in a relationship and felt like you were the mother? Have you ever played the role of the caregiver? Have you ever been in a relationship where you have felt like you were her father? Let me help you with something. It's not their fault that you played the role of the mother. It's not her fault that you were like a father to her. Those were your own personal elephants that you have been projecting onto your relationship and blaming somebody else for. That's something that perhaps reflects your own childhood. Nobody made you assume those roles. You willingly accepted them.

6. Low Self-Esteem

Your parents or your past relationships often degraded you and they never gave you compliments and told you that you were beautiful or handsome. So now you expect for somebody to overcompensate for what others did to you. You enter into relationships with low self-esteem and you place the onus on somebody else to make you feel good about yourself. Now in your relationship, you count how many times the other person says, "I love you" or how many times he tells you that you are beautiful. If they don't say it enough, you hold them hostage, or when somebody else pays you an empty compliment, you think that they love you and that you were meant to be together. So you jump in head-first simply because you have issues in your past that you have not dealt with and have not been healed from. You leave others to wrestle with your elephants because you choose not to address them yourself.

The thing about elephants is, for some reason I can see your elephants easier than I can see my own. I know it's hard to believe, but when it comes to viewing somebody else's unresolved issues, it's like looking out of one end of the binoculars. That is, until it's time to assess your own issues, we then flip the lens. We can't see our elephants because we simply choose not to put the same effort in discovering and dealing with our issues as we do other's. There's a 90% chance that your relationship failed not because of infidelity, mismanagement of finances, bad sex, or the lack of communication. Those all may very well be valid complaints from your past relationships, but they are all symptoms of the problem, not the cause. The reason your past relationship suffered is because there were unresolved emotional issues present that had not been called out and dealt with. As a result, you've spent your entire relationship taming, training, and dancing with your partner's elephants and not enough time with your partner.

Many relationship experts today will tell you the number one reason relationships fail to work and marriages end in divorce is because of these unresolved emotional issues of our past that sabotage our present. We repeat, replay, and project onto our partners unfinished business from our former relationships, and many times, our childhood. Peter Michaelson, a psychotherapist, suggests in *The Dire Detriments of Divorce* there is an unconscious program within us by which we learn to collect injustices from our past. When circumstances or behaviors in our current relationship remind us of these old emotions and grievances, negative transference and projection kick in. Our inventory of injustices is unconsciously released upon our partners, undeservingly.

Section Two: RECYCLING BAD HABITS

Many relationships and marriages falter due to one or both people transferring their partner into roles and characters they were never meant to be.

Unsettled business with parents, past hurts, disappointments, and unmet expectations of previous relationships can often get misplaced onto the person you are with. What is worse, ignorance, arrogance, and pride typically prevent a person from even being willing to admit that elephants exist in the relationship. They don't address the unsettled issues and hurts of their past.

> Many relationships and marriages falter due to one or both people transferring their partner into roles and characters they were never meant to be.

I recently conducted a relationship seminar in Alabama where I discussed the idea of 'elephants in our relationships,' and our bad habit of transferring them into our relationships, when a woman engaged me. She was about 62 years old and she began to confess to me how ever since her youth she has tried to fight for her father's attention, affirmation, and approval, but she never got it. She acknowledged that she carried elephants into her relationships her entire life expecting the other man to tame them and to tolerate them. She confessed that for a long time she tried to hide her elephants underneath a blanket but they no longer wanted to remain hidden from plain sight. She, as a result of this revelation, decided for the first time in her life to deal with her own elephants and to remain single until she did.

When you don't deal with your elephants, they will find you. There are two things to remember when attempting to deal with the elephants in your visibility. As Matthew 7:4-5 reminds us, you should first, deal with your own elephants before you try to deal with somebody else's. Second, deal with YOUR elephants MORE than you deal with somebody else's.

HEALED PEOPLE HEAL PEOPLE

If hurt people are going to learn how to heal people, then you are going to have to learn five simple principles:

1. Be completely honest with yourself.
As humans, we all have the tendency to downplay our flaws and problems. As a result, we create blind spots in our lives that ultimately cause others harm without us ever realizing the magnitude of the hurt we inflict upon them. Blind spots are real. They can exist without our ever really knowing. If you are going to help heal other people, you must first be willing to face your own blind spots. What are the things in your life you keep hiding from? In order to face your blind spots you need to be prepared to be completely honest with yourself. You have to be willing to dig deep and to ask yourself tough questions. Why am I so clingy? Where does it come from? Am I trying to overcompensate for what I didn't receive in recent relationships? Why am I afraid of commitment? Why am I so insecure about myself? How have I learned to cope with the pain of my past and childhood disappointments? Call it out. Confess it, and stop suppressing it. Sometimes it might help to hash these things out with an honest and trustworthy friend who has your best interests at heart and who will not agree with you just because you are friends. In order to be in a position to help heal others, you must be flagrantly honest with yourself.

2. It's never too late to get healed.

We will find any excuse not to address our issues. Many times we become so comfortable and complacent in our ways that we convince ourselves, "This is how I have always been." If a 62-year-old woman can approach me and confess that she is in the process of dealing with issues that have plagued her life since her childhood, you can do the same. It's never too late to become a better and stronger person. It's never too late to allow God to heal you.

3. Stop blaming others.

It is in our nature to blame others instead of accepting our own responsibilities. I've wasted many years trying to blame the absence of my father in my childhood as the reason for my temper and my personality flaws. I've tried to blame my mother's alcoholic abuse as the reason I feel the need to be a hero, today. Truth is, the only way I can unpack my baggage and unravel the ball of confusion in my own life is to learn to own and embrace the issues that belong to me. You can't punish the people in your present for things that happened in your past when they had nothing to do with what happened. You can't continue to blame the absence of your father as the reason you have daddy issues or a problem trusting others. At some point in your life, you must take responsibility for your actions. Whatever happened or whoever let you down in your past should not be a problem for somebody in your present. No one else should have to suffer because of what somebody else did to you.

4. Forgive people.

Stop holding your past hostage and begin embracing the freedom to forgive. Forgiven people forgive other people. Free people learn to free other people. You cannot be any good for anybody else if you are still holding on to unforgiveness in your heart for other people. You can't be fully healed by holding onto hurt. To forgive somebody is to free yourself. "Freeing yourself was one thing, claiming ownership of that freed self was another" Toni Morrison.

5. Refrain from dating too soon.

Are you really ready to date, again? Are you in a position to fully engage a relationship without distraction? Are you emotionally available? You help to heal people when you have the capacity to keep from hurting them unnecessarily. When you avoid your impulse to date (knowing you don't have the emotional capacity to carry a relationship at this time in your life), you choose the best path. When we date too soon, we run the risk of choosing unhealthy partners or we allow ourselves to be unhealthy for somebody else.

CHAPTER 7

*AVOIDING THE BEHAVIORS THAT
SABOTAGE YOUR DATING RELATIONSHIPS*

SEVEN HABITS THAT MAKE GOOD RELATIONSHIPS GO BAD

1 **You make your relationship a rescue mission.**

When we commit to a person, we make a commitment to accept them for who they are. This includes, their personality, their strengths, their shortcomings, and their past. Our purpose for committing is not so that we can open the toolbox and begin our lifelong mission to make Mr. or Mrs. Perfect. Relationships were not designed for us to change all the things about that person we realize we don't like. T. D. Jakes once said, "A man shouldn't have to save you to love you." You shouldn't have to recreate a person in order to commit to them. If you feel that you have to be the rescuer or that you need to save the other person in the relationship, chances are that relationship isn't for you. In the end, you become emotionally exhausted, and in the process, you or both of you grow resentment towards one another. I know. I've long called myself "the firefighter" because I grew accustomed to the thrill of capturing, rescuing, and improving the women I dated. When my ex-fiancé first called me a rescuer, I stuck my chest out, because, in this context, it was good. Well, at the time, we both thought it was good but

> Our purpose for committing is not so that we can open the toolbox and begin our lifelong mission to make Mr. or Mrs. Perfect. Relationships were not designed for us to change all the things about that person we realize we don't like.

it really wasn't. When we obsess over trying to fix people, it becomes very revealing that we have unresolved issues from our past and perhaps from our childhood, which continue to sabotage our present.

In a conversation between Lucy and Charlie Brown: Lucy said, "I think I want to change the world." Charlie Brown responded, "Well, Lucy, that's awesome. Whom would you start with?" Lucy replied, "You, Charlie Brown. I would change you, first."

Your relationship is not designed to be a rescue mission. Your partner is not some grade school science project. Upgrading your boyfriend is not your objective or an accomplishment you should seek to undertake. There are no awards for fixing people, there are no envelopes to open with your name on it, and there are no trophies. You don't work for Olivia Pope and Associates and your name is not Tim "The Tool Man" Taylor. If you are in a relationship with somebody that you feel you must "fix" in order to fully love, then do yourself a favor, let them go. When your relationship becomes more like a home improvement plan, you should rethink your priorities. They will be happy with somebody else that will love them for who they are and the quirks they may have.

> NR
> When your relationship becomes more like a home improvement plan, you should rethink your priorities.

2. You gave him the entree when he only had enough to buy an appetizer.

Sometimes you can want a relationship and a connection so badly that you give up too much, way too soon. We live in a culture where you either make the appetizer the main entree or you will skip the appetizer altogether and only order the entree. Many times this takes place because most people either don't have the money to spend or they only have an appetite for one and not the other. If you are dating somebody who doesn't have the capacity to take their time getting to know you better, or if they only have an appetite for sex, then you need to save yourself the heartache and walk away. Everybody who sits down at your table to eat doesn't deserve your food. You can't keep giving up your entree when the other person is only interested in purchasing an appetizer or sampling your platter. Like I have said before, if you value your worth you will stop giving people discounts. You need boundaries. It sounds cliché, but it is true. When you're too accessible, too available, and your "No" is rusty from remaining unused for so long, you relinquish your influence and power over to the other person. When a man has access to you, on-demand, whenever he desires, it is a higher probability that he will not appreciate and respect you fully. I know it sounds trivial, but when you develop boundaries, draw lines, and ration out your time and attention, he, in turn, will learn that you respect yourself and you require that he follow suit. Respect comes before love, and boundaries help to establish respect. We easily sabotage our relationships because we throw caution to the wind and do not establish rules regarding how you prefer your relationship to unfold and how you prefer to be treated in the relationship. Boundaries do two things: they keep things and people from coming inside and they keep things from the inside getting on the outside. They are not designed to imprison you but to protect you. We fail to create boundaries in dating and instead

> NR
> Everybody who sits down at your table to eat doesn't deserve your food. You can't keep giving up your entree when the other person is only interested in purchasing an appetizer or sampling your platter.

invite behaviors inside our relationships while it's still in its infancy. This forces our relationships to age rapidly, like the Curious Case of Benjamin Button. We date and yet our relationship looks more like a marriage just without a certificate.

3. You Press The Play Button On Your Relationship, The Pause Button On Your Life, and the Stop Button On God.

It happens more than you and I want to admit. You meet a person that you like, and, as a result, you want to impress them. When you commit yourself to a person you want to make sure that they are happy. You want them to feel like they would rather be with nobody else in the world but you. Unconsciously you find yourself always thinking about them and how to make the relationships better. You spend time, money, and energy trying to be the perfect partner. This is perfectly fine but while you are focusing on making your life with your partner the best, many times, you inadvertently put your needs and priorities on hold. In most cases it is unconscious; we start dating and then we stop everything else. We shrink and reduce our own interests and values for the greater good of the relationship and we concentrate and amplify the life and happiness of our partner instead.

> **NR**
> You may not know this, but when you begin dating, your life should not have to take a backseat. You must continue to feed your dreams, nurture your passions, and grow toward spiritual maturity.

You may not know this, but when you begin dating, your life should not have to take a backseat. You must continue to feed your dreams, nurture your passions, and grow toward spiritual maturity. You cannot live in the shadow of your partner. You are born with an identity that needs to be separate from anybody else you date. Submerging yourself in the life of your relationship without giving heed to your own life can have dire consequences. When you press the play button on your relationship you should also be pressing the pray button, too. Try not to begin your relationship in a frenzy. Create a healthy balance between you and your partner's needs.

4. You race against the clock.

Some say Danny Biasone saved the NBA when he invented the 24-second shot clock rule. The purpose was to speed up the game and to make it more exciting. The offensive team has 24 seconds to shoot a ball that hits the rim but if they fail to do so they will be in violation and have to surrender possession of the ball. As a result, coaches scheme plays and teams develop ways to play efficiently as they race against the shot clock. Teams who underperform typically mismanage their time on the court because they are not maximizing every moment they have when they are in possession of the ball. In relationships, we have created a clock that forces us to speed up the dating game and to fast forward relationships. As a result, if we don't get married, have children, and begin a family by a certain time in our lives we will be in violation and consequently be branded as "strange" or "defective." Tick tock goes the biological clock. There seems to be this fear that time is running out. Sure, we all hear it ticking within us: graduate college by 21, begin lucrative career by 23, fall in love and get married by 25, and begin having children no later than 27. It's the perfect timetable. Unfortunately, this timeline sets the standard for happiness and success in our society. Most women, and even men, find themselves sabotaging their relationships and settling for a fraction of what God intended for them to have, simply because they were racing against the clock. We rush to get married as if marriage is the finish line in relationships. We bypass and overlook the simpler things in life and the many blessings God has placed before us. We do this because we prefer to live in the express check-out lane to be "happy." Your life and your relationship deserve your taking the time to embrace each day that God has given you.

We operate from the belief that our clocks are always ticking yet nobody knows whether "the clock" is on Central Standard Time, Mountain Time, North America, Western Sahara Time Zone in Africa, or Irish Time Zone in Europe. From which of the 24 time zones does this imaginary clock operate? When you race against the invisible clock, you overlook the gift of the present and you allow your success to be weighed by society's standards and not God's.

There are two different time zones you can choose to set your clock to follow, either God's Time Zone, Kairos, or Man's Time Zone, Chronos. Chronos, from which we get the word chronology, is the realm of time we all exist and adjust our watches and clocks to follow. Kairos, a New Testament Greek word, represents the fullness of time and the arena of man's decision on his way to an eternal destiny, according to scholars. Kairos refers specifically to periods of divine determination. Since God created time, He chooses to enter into Chronos time according to His perfect will and create what some call a Kairos moment. A Kairos moment is when God's timing intersects Man's timing. You must be willing to allow yourself to operate from God's timing and not your own. We pray for a Kairos moment and to stop allowing our clocks and agendas to interfere with God's timing, plan, and purpose.

5. You forget to check your emotional baggage at the door.

I love to travel. I love to see new things and visit new places. It's what life is about. However, there are two things that pain me about traveling: claiming your baggage at the airport and unpacking your luggage when you return home. There are those who take nothing more than a "carry on" for their long trip, while others pack as if they were going to hibernate for the winter. The baggage claim carousel is one of the most revealing places you can learn about any person. The carousel hosts cargo of all kind, some small, delicate suitcases and some large, worn vessels. You see people connected to baggage that fits them and others that you wouldn't think belonged to certain people. Some new and colorful and others dirty and busted. Then there is my suitcase, barely holding together. The wheel is loose and the zipper is off track. It's traveled so many places and has experienced so many weather conditions. I always say to myself "When I get time" I am going to get me some better looking luggage but I never really get around to prioritizing that on my to-do list.

The result of traveling, however, is that I have to pack luggage and unpack it when I return home from any trip. This is exactly what I fail to do. The packing up is easy, it's the unpacking that I don't do so well. The most time consuming thing, for me, is to unravel all of the dirty clothes and sort through socks and shirts to determine what gets washed and what goes in the clean clothes drawer. Because I have a tendency to accumulate new things on my trips, I have to segregate my new things from my old things, too. Although I have the greatest intentions, I usually don't get around to unpacking my baggage, not until days and sometimes weeks later. It's totally embarrassing. One time it got so bad that I fully operated from my suitcase for a week.

Now picture the relationships you have engaged over a lifetime as traveling and your life's experiences as the emotional cargo piled up over time. What does your suitcase look like? Are you traveling with "carry-on" luggage or large vast suitcases? Are you "that person" who checked in two huge bags, both of which exceeded the weight limit? What is your process of debriefing a trip, or a relationship that didn't work? In my experience, everybody has emotional baggage and we take that into every aspect of our lives. So much of our life's experiences pile up. Instead of taking the time to deal with our baggage, we carry them into our next trip, or relationship. We bury what we perceive to be hidden emotional treasure into small compartments of our suitcases and leave them as an unpleasant surprise for the next traveler to find. Parents, family, past relationships, abuse, addictions, and many other secrets contribute to the vast amounts of emotional baggage we try to tuck away so we never have to address or deal with them. When you try to suppress and avoid managing your emotions, they surface in ways that can critically handicap your relationships. When you have baggage from an old relationship that continues to show its face, it contributes to an emotional absence in any relationship. You need to address it because an emotionally unavailable person is the most frustrating person to date.

> **NR**
> When you try to suppress and avoid managing your emotions, they surface in ways that can critically handicap your relationships.

6. You have unexpressed and unrealistic expectations.

One of the biggest and most common mistakes you can make in a relationship is to pretend you don't want anything serious but continue

Section Two: RECYCLING BAD HABITS

to casually sleep with him or her in hopes that things will grow beyond the physical relationship. Unexpressed expectations are the cause for collapse in most failed relationships. Are you clearly communicating your needs in a way she can comprehend? Are you communicating your feelings in a way he will be able to translate? I heard somebody say, "You can't work from a shared perspective unless you share your perspective."

When you don't clearly express and communicate your expectations it isn't fair to expect the other person to respect them or to meet them. What do you need to be happy in your relationship? What do you expect from him that will allow and encourage the best presentation of yourself in the relationship? How do you really feel about the state of your relationship? You owe it to yourself to speak up and to be clear about what you desire. In many cases, we don't create a safe culture in our relationship where we can exchange vulnerable dialogue without the fear of punishment of some sort. As a result, we accumulate frustration and an expectation gap and ultimately find other things and people to which we feel we can release them safely. This is where the divide happens. When you allow unspoken expectations to fester they, will eventually surface in some way.

> In many cases, we don't create a safe culture in our relationship where we can exchange vulnerable dialogue without the fear of punishment of some sort. As a result, we accumulate frustration and an expectation gap and ultimately find other things and people to which we feel we can release them safely.

However, we must be careful and cautious that our expectations are real and not ideal. We sabotage our relationships not only by unspoken assumptions but with expectations that are unrealistic. Studies show that it's more likely that relationships will develop problems when expectations are unreasonable. We all have a preconceived notion of what we want in our ideal mate, this is how unrealistic expectations are generated. We project our childhood experiences, past relationships, and family traditions onto the other person and how we want them to be. Expectations build up over a lifetime of experiences. It's not the other person's responsibility to fill the void left by somebody else, or to give attention to an area of your life that needs your undivided attention. We must be careful that in communicating our expectations to our mate, they are clear, they are reasonable, and they are not misguided. Having unrealistic and unexpressed expectations is a sure way to ruin a relationship.

7. You don't define the relationship.

Sure, you're having a great time together a few times a week or several times a month, but do you really know where you stand? Are you in an open relationship? Exclusive? Dating? Just Friends? Is one of you hoping it will turn into marriage and kids while the other is a commitment-phobe that enjoys seeing more than one person at a time? After a few dates, sit down to discuss your thoughts on relationships, commitment, and how you'd define where you currently are — and where you might be headed.

I know it seems elementary and infinitesimal, but so many relationships are compromised because people fail to define them. When your relationship goes undefined, accountability gets lost. Nobody defined "dating" or "relationship" and yet you assume the other person is operating from the same perspective, when, in most cases, they do not see things the way you see them. I had a friend tell me that she recently met a man who eventually asked her out on a date. After a couple of weeks of getting acquainted with one another, they had dinner, went to a movie, and later returned back to his house. They crossed certain boundaries that night. She told me that the week following this encounter, she noticed he had stopped calling her. When she confronted him, he told her that they were just "hanging out" and that they never went on a date, nor was he looking to date her or become serious. She expressed disappointment and confusion because she assumed they both were on the same page and they wanted the same thing. She couldn't have been further from the truth.

It is imperative to define dating and what you both are doing at the onset of seeing one another. How one defines and understands this term can make all the difference in how the relationship unfolds. There are many views and opinions about dating and committed relationships and yet there remains very little meaningful conversation about it among couples. My view versus your understanding can determine the trajectory of our relationship. It can be a matter of two different destinations.

It's kind of like two people profusely excited about going on a trip with one another away from home and yet, without having significant and intentional discussion about where they want to go, how they want to get there, and what they hope to do when they arrive to said destination. They get into the vehicle and immediately jump on the road and the person driving has on his mind a local hotel right outside the city while the person in the passenger seat can't wait to get to the beach resort in Florida so she can wear her bathing suit and get a much needed tan. They are both excited. They both want to go on a trip together. The only problem is, essentially, both people have two different destinations in mind and because they have learned to operate from assumptions, it heightens the probability of a disappointing experience. They are both going somewhere, but the problem is that they are going in two different directions. They have two different expectations. Don't allow yourself to get so caught up in the excitement and newness of any potential relationship that you forget to discuss the most important matter, your destination.

> **NR**
> Don't allow yourself to get so caught up in the excitement and newness of any potential relationship that you forget to discuss the most important matter, your destination.

CLOSING THOUGHT

"Have patience. God is not finished with me yet."
Philippians 1:6

Pray this prayer:

God, although I am not where I desire, I am so glad that I am not where I used to be. My life is a construction site, and You, God, are still working on me. Help me to abandon the destructive habits that are keeping me from progressing. Help me to focus on becoming the best version of myself before trying to create a life with someone else. And when the time is right, help me to develop covenant eyes so that I am able to identify and engage healthy people who will help to foster healthy relationships.

Amen.

Section 3
GETTING PAST YOUR PAST

"Forgiving does not erase a bitter past. A healed memory is not a deleted memory. Instead, forgiving what we cannot forget creates a new way to remember. We change the memory of our past into a hope for our future."
-Lewis B. Smedes

CHAPTER 8

OPENING YOUR HURT LOCKER

WHEN THE PAIN OF YOUR PAST DISRUPTS THE POTENTIAL OF YOUR PRESENT

Dear ex-husband, my heart is still bleeding from the betrayal you caused me. Dear uncle, the vivid imagery of how you took advantage of me still plays through the windows of my mental enterprise and cripples my ability to love someone fully. Dear wife, you've been estranged for some time now. You promised me things would be better but they're not, you're still loving somebody else. Dear ex, your lies still reverberate through my consciousness, I can't trust anymore. Dear mom, you abandoned me. Look at me now, I'm a flight risk to my own children and my family just like you. Dear self, how long are you going to keep running and creating hiding places to bury your emotions and cover up your pain? When is it going to stop? The blame-game? The self-infliction? Running from one set of arms to another?

There are many of you who have experienced betrayal that runs so deep one can get lost in the abyss. And yet, it's easier for others to say, "Forgive him or her, and move on." Moving forward from any broken relationship is by far the hardest thing any one person can do. It's not easy. There are remnants left behind, and, as a result, they've been lodged within the very crevices of your heart, mind, and spirit. What is worse, these shrapnels camouflage themselves and give you the illusion that you are better, you are healed, and your past is in the past. When perhaps it may be the furthest from the truth.

Everybody has hiding spaces. Small orifices in our lives where we tuck away the things we consider meaningless but they obviously have enough meaning because we hide them away rather than deal with them. Unfinished business doesn't go away. It keeps recurring, replaying, reappearing and sabotaging our lives until we choose to stop and deal with it. You must create for yourself a safe environment, and slowly and meticulously begin to open yourself up to your pain and your hurt. This is something you perhaps have never been privy to. You have to open your locker that hides and protects all the artifacts garnered from your past hurt. Deal with them. Nobody wins as long as you choose to keep silent. David said, "When I kept silent, my bones grew old through my groaning all the day long" (Psalms 32:3).

CLOSE THE EMOTIONAL BACK DOOR

[journal entry]: Reappearing
My talent. My way with words. My charisma couldn't save me this time. She confirmed my heart, the destruction site, as a state of emergency; a disaster zone. You can have a good thing but if the timing is bad, then, it remains to be a bad thing. You know, in all my life, in all my dating, there was one that I let get away. We all have that one. Her name was Brittany Spears. She had my heart. I loved her like no other. We had something special. When I was around her, all I could utter was, "I want to marry you." It was just that bad. When she would say jump, I would say how high? The Lord seemingly allowed us to reunite this year. I was thrilled. I thought to myself, "It's time." I was in a very peculiar season in my life. God was dealing with me on a personal and spiritual level. My world was in an emotional uproar when she and I reunited, when she reappeared in my life.

Although I knew I had the capacity to get distracted by her, I really wanted it to work this time. My only prayer to God was, "Is this a distraction?" Really, I didn't want to pray because I didn't want to risk the Lord saying something I didn't want to hear. But I prayed anyway. I heard and felt the inevitable - "Not now. Trust me and wait. I will give you the desires of your heart." I took it

Section Three: GETTING PAST YOUR PAST

> *upon myself to pursue the relationship anyway. Slowly after I decided to follow my will, things began to crumble. My past came back to haunt me. I told her a lie ten years prior that haunted her. It haunted the both of us. It wasn't until now that I wanted to confess and release myself of it. I wanted nothing to stand between us. She couldn't get past it and I understand why. I was a liar. I knew the Lord was telling me to slow down because I had just come from one emotional roller coaster to this one, it was a red-eye, nonstop flight. I couldn't understand for the life of me why the Lord would allow this to happen to me. Maybe I deserved it. Actually, I did deserve it. This was a period in my life that my heart was broken into a million pieces. For all the pain I've ever caused anyone, this one's for you. She opened my nose wide, spread my heart open, only to hit me with a dagger straight in my chest. Boom. One million pieces.*
>
> *Nothing is worse than to have your past catch up with you. To haunt you. To come creeping out of your closet. The only way I can make sense of this is that maybe God wanted me to clear my heart and my slate clean from anything I may have thought needed to be there. My sin cost me the one thing I thought I ever wanted and that was to be with her. That door is no longer open; it's closed. There is no door anymore.*

Picture a door. On one side of the door, it's cold and gloomy and dark. This is where you used to be. This is where you once used to live. This is where you are trying to leave. There is pain and heartache on the outside of that door. Memories of what used to be, sit idle on the other side of that door. There are also harsh realities that you conveniently misplace as you try to hold desperately to those memories on the outside of that door. Don't forget. Remember those bad habits you developed. Remember the depth of the pain you experienced when you were on that side of the door. Some of us lost ourselves when we were on the other side of the door. We lost control of the relationship. Things went awry. Remember, you thought you would never survive. You thought you would not get out alive. You didn't know what tomorrow would provide for you. You thought it would always be cold, gloomy, and dark. People always told us, "This too shall pass." We couldn't see it, though. We didn't think we would make it past that last heartbreak. We didn't think we would be able to recover from the selfishness that destroyed our relationship. Yes, there are some things we regret, but we are no longer on that side of the door. God's mercy and grace allowed you to experience the other side of the door.

On this side of the door, it is warm, bright, and peaceful. You don't have to fear for your life anymore. You have peace. This peace isn't coming from another person; you are finding peace with and in God. You aren't questioning your self worth. You are now building the blocks again to restore your life. This door separates us from that person or those people we love dearly. The more we get acclimated to the warmth on the inside, the more we consider our life with the other person on the outside. So why would you ever want to open that door again? Why would you want to leave the door cracked open? We find ourselves severely tempted to unlock that door and to twist the knob to let them in so they can experience and witness the life we have built for ourselves without them. The truth is, we want more. We want to resuscitate the life we once had with them. Then we somehow notice that the temperature on the inside of the door to our home is dropping. But why? The cold from the outside has seeped inside because we left the emotional door to our past relationship cracked. You begin to lose the warmth and you compromise the safety of everything you're building on the inside when you leave the door of your past on the outside open.

"Maybe, they will decide to come on the inside and choose to live the life you desire for them. Maybe if you give them another chance they will act right this time." You try to convince yourself.

You forgot, that same line of reasoning didn't work the last time so why would it work now? The inconvenient truth is, to leave the door open is to risk sabotaging your life and your sanity. You can't just hope things will get better because hope doesn't keep you healthy and whole. Leaving the door open threatens your safety and your peace of mind. You can't effectively build upon your house because you are spending too much time looking through the peep-hole of your past through your rear door. God is trying to send that special somebody to your front door but you can't answer it because you are having such a hard time closing your back door. You can't fully entertain the company you have on the inside with you because you are too busy meddling with the back door. God allowed you to find your way to the inside of the door not so you can continue to flirt with returning to the cold on the outside. Your emotions and memories on the outside need to remain there. Some things cannot be recovered. It is the cost of dating.

How do you close the back door? You close the back door by removing your mask, having a yard sale, and releasing yourself.

REMOVE THE MASK

Writing this book has been particularly difficult for me. I realize that even as a preacher, a man, and a leader, I have so much that is pent up within me. My spirit is full, emotions on overload, and I have blind spots that have hindered me from being able to operate in any relationship at full capacity. My frailties and shortcomings have become such a normalcy that I have learned to operate optimally at half guard, sort of like a functional alcoholic, just without the alcohol.

I have this old white Jeep Cherokee that I am still driving from when I was in college. With 200,000 miles on the motor, my truck naturally needs maintenance more frequently. Embarrassingly, I have a check engine and airbag light that has been on for nearly one year now. Not once have I taken my truck to the mechanic to get the problems addressed. I keep saying, "I'll get around to it." Finances, time, priorities, and complacency, all kept me from getting my vehicle fixed. I have grown accustomed to operating and driving my vehicle, and worse, carrying other passengers in it despite this truth. As long as my truck cranks, I would drive it.

As I began writing this book I was forced to take a hard look at myself in the mirror. God began to deal with me in a way that He hasn't been able to deal with me before. Like my truck, I was in dire need of maintenance. The condition of my truck was indicative of the condition of my heart, they both were operating under high mileage. I grew accustomed to ignoring the wear-and-tear of a weary heart. As long as there was a passenger ready and willing to enter a romantic relationship with me, I would oblige. They say, "life goes on" so I tried to "go on." After my failed engagement, I began dating again. More like, soon, thereafter. I confess. I was guilty of the "Big Over and Under." In order to get over somebody, I tried to get under somebody else.

During the course of this new relationship, I inadvertently erected an invisible barrier between her and me from which I learned to function and to hide certain emotions. It was a new normal. Regret festered in my spirit. I covered things up. I ignored truth. I was harboring a hurt locker of stowaways. I had treasures, broken promises, and other memorabilia that still needed to come out. I was reluctant to listen to my heart telling me it needed maintenance. I thought it would be in my best interest to remain in the relationship instead of risking another broken heart all because I needed to take time for myself. I know it sounds ridiculous, but why hurt somebody else? Guilt kept me bound. She didn't deserve it but I had never fully given my heart to this woman. I was only partially visible to her and she didn't realize it. It was sort of like I was wearing a mask in my relationship.

Professional wrestlers wear masks in entertainment to disguise their true identity and to entertain the public. Doctors wear masks in the medical field to protect themselves from sickness and disease and to help save the lives of their patients. Hockey goaltenders wear masks in sports to protect their heads from injury in a fierce game that is played on ice. These and so many others are required to mask their face in order to excel in their profession. However, in relationships, wearing a mask slowly kills the relationship, and it's everything but right.

> In relationships, wearing a mask slowly kills the relationship.

I grew accustomed to wearing a mask and I needed to stop the masquerading, so I decided that I had to break up with the woman I had been dating. I thought it was because I just needed the space to sort through our current dating situation and the problems we were having. I later realized it was because my heart was still not healed from my relationship nearly three years prior. I had to admit to myself that I was still hurting from my broken engagement. Remnants of an improvised explosive device (IED) was lodged in my heart and threatening to destroy my life. I needed to stop masquerading and acting as if I were self-sufficient, healed, and whole and admit that I was in desperate need to have God maintenance my heart. Breaking up with her was very difficult. It made me sick to my stomach. But because I unconsciously left the back door of my past cracked, I needed to secure all the entrances to my home. I needed to close my back door.

An excerpt from my journal:

> "...Today, I broke yet another heart. Today, I closed the door in the face of yet another beautiful, promising woman. Today, I stand hurt, confused, disappointed, and angry that I held the heart of a woman in my hands and chose to give it back to her. I promise, I am so sick of hurting myself and hurting others that it practically kills me to even think of dating and marriage. I thought I dealt with my past? I thought I situated things. Obviously not. God, somebody is going to get hurt today and I feel terrible about it. I am a horrible, horrible person. The more I try to lead by example, the more I mess up. My God, forgive me for I have sinned. I receive no honor in breaking another heart. Jesus."

Are you wearing a mask? Is your heart functioning under high mileage? Do you have a hurt locker of stowaway treasure that needs to come out? You need to stop and make yourself priority.

YARD SALE

I know hate is a strong word but I hate moving. I hate everything about it. Why? In my adolescent-hood we moved at least 28 times in the span of about 10 years and along the way, we accumulated "stuff." Some things we needed, others we kept simply because it was easier to keep them than to throw them away. I went off to college with furniture given to me by my mother that we had for nearly all my childhood years. The couches were old, worn, and useless but it was how we had learned to survive. Every year during tax return season my mother would go out and find new things and newer knick-knacks to place in our home. She would do this, rarely getting rid of any old items. We accumulated a lot of unnecessary things, so when it came

Section Three: GETTING PAST YOUR PAST

time to move, it was that much more excruciating. When I think of moving and toting and packing and carrying furniture and belongings, it literally makes me nauseous.

As an adult, I treated my emotions and my relationships the same way I had learned to treat furniture while I was growing up. In my clumsy rummaging around looking for love, I discovered that both my heart and mind were full of clutter from my past. The only way I was going to be able to move forward, was to embrace the reality that I must have a "yard sale".

> You have made it virtually impossible for God to bless you with a future because you are holding on too tightly to the things of your past.

We need to have a yard sale emotionally and spiritually because many of us have too many things in our lives that are taking up too much space. Our minds, our hearts, and our lives have become far too cluttered to be able to move forward and receive the blessings He has for us. Many of us have too many things and tokens from our past that are taking up unnecessary space in our thoughts, occupying our minds, and too many broken things that no longer work. They are still lodged in the crevices of our hearts simply because we don't want to get rid of them. As a result, we cannot redecorate our lives and be renewed from our past because there is not enough room. You have made it virtually impossible for God to bless you with a future because you are holding on too tightly to the things of your past. A yard sale is likely the most rational thing to do. Why? Because somebody else would like to have the things and people you're holding onto. Your "trash" would be somebody else's treasure if you would just release him, or release her. Just because the relationship was toxic for you doesn't mean it will be toxic for somebody else. Release that person to God for somebody else. Let them go. It has become too cluttered in your life to be able to truly function the way you are supposed to. I admit, sometimes it is difficult to throw away, discard or give away things that aren't broken, even if they are only in pretty good condition. You have to ask yourself whether or not this relationship has run its course. Is this relationship the place God has for you? Are you both able to invest in and still flourish while in one another's life? Is this the spouse that God has ordained for your life? It may be that she or he was designed to be in your life for a season but you have become so selfish and shortsighted that you try to keep carrying this person into your future with you. Perhaps their season has passed.

> Having a yard sale is sometimes tedious, inconvenient, and time consuming but when you finally open your doors you begin to recognize all the things you've been holding onto.

The difficult thing is learning to let go. Letting go is scary. You don't know if you will ever come close to having something that will at least generate some type of emotion within you. You don't know if, when you let this person go, somebody else will find extreme happiness in them and you will be left alone without somebody to love you. You fear letting go because there is so much unknown. This is the risk of letting go. In essence, you are operating in faith that in due time, God will supply your every need and give you the desires of your heart. Having a yard sale is sometimes tedious, inconvenient, and time consuming but when you finally open your doors you begin to recognize all the things you've been holding onto. You learn that now somebody else will take value in them, and this is okay.

Jesus, help me to have a yard sale, please. Help me to identify, discard, and release those things in my life that are now taking up space in my mind, my spirit, and are not helping me to be fruitful.

FRIENDLY FIRE

In warfare, the greatest loss any military constituent will ever suffer is injury or defeat at the hands of their own forces. This is otherwise known as friendly fire. The single most debilitating thing about combat is becoming the target of an attack that was originally intended for the opposing forces. In the wake of the second Gulf War, the shooting of two U.S. Army Black Hawk helicopters by USAF fighters during the Iraqi no-fly zones caused for a terrible experience with friendly fire. The advanced technology currently used to reduce friendly fire aboard the helicopters failed causing them not to show up as friendly forces to other military personnel. The US Air Force investigation placed blame on many factors. One of which was the failure of the Identification Friend or Foe, or IFF systems aboard the plane. The other was the misidentification of the aircraft by other forces.

In dating, relationships, marriage, and divorce, the greatest loss any person will ever suffer is injury or defeat at the hands of oneself, or friendly fire. In our attempts to hold grudges, harbor anger, and enact revenge upon our offenders and those who have misused us, we fail to realize that the only person who really suffers is ourselves. Unforgiveness is a form of friendly fire. When we hold onto past mistakes, hurt, anger, and regret, our Identification Friend or Foe, or IFF systems, malfunction. Sickness, depression, bitterness, and even death, are often created and nourished by unforgiveness. Unforgiveness has no destination. It never proceeds in the direction we intend for it to go. It's predictably unpredictable. When our anger festers, our own body suffers. Our thoughts and consciousness become polluted, all the while, the person or people we have tension with are free from burden. Truthfully, some people don't deserve your forgiveness. Even I sometimes don't feel like forgiving people. Have you ever just wanted to be angry with somebody simply because they deserved it?

What is one extreme violation you have experienced in your life by somebody else? How did you handle it? Have you released yourself? Are you still bound by it? I haven't walked a mile in your shoes so I can only imagine. Truth is, we don't deserve the forgiveness from God we have received. The thing about forgiveness is you have to forgive people even when they don't deserve it. According to Lewis B. Smedes, forgiveness is about letting a person out of prison only to realize the person was you. Although you don't feel or want to forgive your debtors, you must, because in all honesty, everybody deserves forgiveness. Everybody has a right to have an opportunity to encounter the grace of God, and sometimes the only God people will ever encounter will be through you.

David's ability to forgive Saul, his public enemy number one, didn't come easy.

"So David left Gath and escaped to the cave of Adullam. Soon his brothers and all his other relatives joined him there. Then others began coming - men who were in trouble or in debt or who were just discontented - until David became the leader of 400 of them." I Samuel 22:1-2, NLT

"So David and his men - about 600 of them now - left Keilah and began roaming the countryside. Word soon reached Saul that David escaped." David now stayed

in the strongholds of the wilderness and in the hill country of Ziph. Saul hunted him day after day, but God didn't let Saul find him." I Samuel 23:13-14, NLT

"When David heard that Saul and his men were searching for him, he went even farther into the wildernessss to the great rock and he remained there in the wilderness of Maon. But Saul kept after him in the wilderness." I Samuel 13:25, NLT

"At the place where the road passes some sheepfolds, Saul went into a cave to relieve himself. but as it happened, David and his men were hiding farther back in that very cave. 'Now's your opportunity' David's men whispered to him. 'Today the Lord is telling you,' 'I will certainly put your enemy into your power, to do with as you wish.' So David crept forward and cut off a piece of the hem of Saul's robe. But then David's conscious began bothering him because he had cut Saul's robe. He said to his men, 'The Lord forbid that I should do this to my lord the king. I shouldn't attack the Lord's anointed one, for the Lord Himself has chosen him.' So David restrained his men and did not let them kill Saul." I Samuel:3-7.

David lost everything, his wife, Michal, his job working in the palace, his best friend Jonathan, and his mentor Saul was trying to kill him. David was brokenhearted and resorted to living his life on the run. His inability to face Saul and to forgive him caused him to retreat into hiding. He was living in dark places, caves. Eventually, David had to come out from hiding in the dark. At some point, David had to exchange his fear, his pain, and his anger for God's power to forgive. Despite whatever you have been taught about forgiveness in your Sunday school class, forgiveness is not natural. Forgiveness is supernatural. Sure, its easy to forgive somebody who steals a pencil from your desk but when somebody steals 10 years of your life through a turbulent marriage, forgiveness doesn't come that easily. If you're wondering why you haven't been able to forgive that certain person in your life, it's probably because you've been trying to do it under your own power.

You don't have the capacity to look beyond a person's faults without the help of the Holy Spirit. Whether you are forgiving somebody else or you are learning how to forgive yourself, you need to be released. When military personnel come under attack and realize they are completely surrounded by the enemy and won't survive the attack, their last resort is to send a distress signal called, Broken Arrow, to their commander. Broken Arrow is a military code word for "direct all fire upon my position." When you attempt to forgive and take on the enemy under your own power, you are walking into a trap that positions you to be overrun by the enemy. When it comes to seeking true forgiveness to those who have wronged you, you need help from your commander, God through Jesus Christ, to achieve it. Send a Broken Arrow distress signal to God through prayer, and with God's help, you will defeat your enemy. The Enemy wants you to harbor unforgiveness so that you can deplete your own self. Ask God to help you experience that release.

> **NR**
> Forgiveness is supernatural.

> **NR**
> The Enemy wants you to harbor unforgiveness so that you can deplete your own self. Ask God to help you experience that release.

Section Three: GETTING PAST YOUR PAST

[journal] The Day I Released Myself From My Ex-Fiancé
August 30 2:07pm

I'm about to send it. I know this is going to sound like I am begging for her to come back to me or that I am weak but I can't focus on those things. I can't control anybody else's thoughts. I must become naked so that I can heal. I need to release myself. My heart is just too heavy.

Dear XXXXX,

"I couldn't help myself. Last night I was looking through photos of you and me. I was looking at the very first photo you took of me. My red shirt. We were in Africa.
At any rate, I am sorry to have bothered you. In all honesty, I have never been able to erase all of our pictures but I have made some great attempts to lock them away in a compartment in my old computer and on a file disk.
I am simply texting you because God has been doing some things in my life - some breaking, crushing, and rebuilding. Through this process I am learning to become more vulnerable and more honest and transparent with myself and the people around me. In order for me to grow closer to God, I must do these things. Here we go, truth is, I am not over you. I am still madly in love with you. I recognize this whenever I come across some random picture of you. I'm not telling you this to vie for your heart or to ask you to reconsider our relationship or to abandon any current relationship you may be in. I'm trying to become more vulnerable, something I have never really been good at. I recognize that I never gave myself an opportunity to grieve over your departure. It's fourteen months later, I never really dealt with it. I just hoped that it would go away. I never could get myself to erase pictures because for me they have so much meaning. We had our differences, but we all choose what we are willing to put up with. I've always been very difficult with embracing vulnerability because for me it was always connected to weakness. This likely contributed to my passive aggressive tendencies that drove you crazy. Anyhow, I am elated to know that you are doing well with yourself. If I graded myself according to society's standards I would probably be considered an under achiever at the moment, but I am okay with that. I am growing spiritually more than ever. The Lord needed to position me where I am in my life so that I would be solely dependent upon him. I am. I am learning to become naked before Him and before whoever else God leads me to embrace vulnerability with. You are one. I apologize for how I contributed to the demise of our relationship. I was emotionally inconsistent. I'm sorry for the hurt I have caused you. I put extreme pressure on myself to be the "man." The one

that was able to provide, protect, and to give you any and everything you desired. I wanted to be the apple of your eye. But all along I was captivated by fear. I secretly feared I would fail you, your parents, and my family. I felt I needed to be further along financially than I was. I wasn't truly trusting God. I was trying to do things myself because God was taking too long. I didn't want to disappoint you. Because of my disobedience, my spiritual immaturity, and my preoccupation with fear, I ultimately failed you, myself, your parents, and most of all - God. I must forgive myself, which is difficult because we had something good. I can only hope and pray that you forgive me. I must release myself from this hidden pain and insecurity of mine. I also must release you. I must now, finally, work on grieving over your departure. I hope you understand. I hope that God gives you all the desires of your heart. I always will be your number one supporter.

This is my hope of a final communication with you. I will always love you."

Issac, your rescuer

CHAPTER 9
CONQUERING THE COMMA

HOW TO IDENTIFY AND RELEASE YOURSELF FROM TOXIC RELATIONSHIPS

Most people don't know this, but I have always had a healthy disdain for English, specifically, grammar, and composition. When I was in grade school, I was horrible with writing. I didn't know much about sentence structure or how to properly utilize punctuations. Believe it or not, it was years later when I was in graduate school that I came to realize I had a problem with run-on sentences, comma splices, and how to effectively use other punctuations. Embarrassing, I know. I didn't know how to use commas, semicolons, and periods, correctly. I learned in graduate school the lessons I was supposed to learn in grade school. I learned that a comma represents a brief pause, a semicolon represents an extended pause, and a period represents a full stop. I had the tendency to put commas in the wrong place, I placed periods in the wrong position, and I significantly abused semicolons. Where there was supposed to be a period to stop any particular sentence, I would not have one; as a result, my sentences would run-on. They sometimes run for days at a time before I caught up with them. Where there was supposed to be a comma to designate a brief pause, I would often place a period instead. It was a pretty awful presentation. I spent so much time in the writing center at Princeton they knew me by name. As a matter of fact, I want to believe they hated to see me coming because for a good while it seemed like I was the only one on campus that just was not making improvement.

You know the saying, "The way you write is a reflection of the way you speak? Through this embarrassing experience, I came to realize that the way I had been writing was also the way I had learned to function in relationships. In my past, I have found that I mastered how to misuse periods because I have brought perfectly good relationships with women to complete halts. Either because of my selfishness, my inconsistency, or because I was just simply immature. I have learned the art of abusing the comma, a comma splice, because of my indecisiveness and my ability to send a perfect message laced with mixed signals.

As a result, I have gone back and forth and played Ping-Pong with women's hearts. I have been in toxic situations where God has tried to place a period in my relationship so that I would stop and let go, but I wouldn't listen. I, in my own selfishness, chose to make run-on sentences of my relationships because I wasn't ready to let go. I remember one relationship so potent in my mind and fresh in my heart; her name was Brittany Spears. I was in a relationship that had expired, and I was fully aware that the expiration date had long since passed. We dated years before when I was in college. We were young, and infidelity on both parts contaminated our relationship. Years later, at a college homecoming, I remember seeing her from afar and my heart skipped a beat. We reconnected right where we had left off years before - in our own dysfunction and inability to remember why we didn't work in the past.

I consciously ignored the proper punctuations required in order to make this relationship work. I knew in my mind it was time to move on but I wasn't ready to let go. My single mission was to make this relationships work. I was more focused on all the reasons it had to work and I avoided submitting to the one reason it would not work. God placed a period in my relationship. It was tough for me but where God had placed a period, I chose to move it out of the way and proceeded to "run-on" in the relationship. I had my own dysfunction down to a science. I was suffering from a toxic disorder called misplaced punctuations. Needless to say, the relationship utterly failed. I wish I knew then what I know now: you cannot manufacture happiness or modify the will of God to meet your expectations in relationships. You must trust in the way God chooses to write your narrative.

MISPLACED PUNCTUATIONS IN YOUR LIFE

Have you ever suffered from misplaced punctuations? Has God ever placed a period in the life of a relationship that you really wanted to work? Have you ever had a relationship you've held onto longer than you were supposed to? Have you ever seen the red flags and ignored the signs because your desire to make the relationship work outweighed your desire to embrace the truth? Have you ever tried to negotiate with reality? Sometimes the most unhealthy situations are the ones we have convinced ourselves we need to remain connected to.

In studying the book of Ecclesiastes, more specifically, chapter 3, I've become fascinated with the grammatical structure and poetic rhythm that flows through the text. It teaches us that there is a natural flow to the different seasons in our lives. It teaches us that no matter where you are, there is still something you must look forward to.

"A time to *kill*, and a time to *heal*; A time to *break down*, and a time to *build up*; A time to *weep*, and a time to *laugh*; A time to *mourn*, and a time to *dance*;" vv. 3-5

But because there is a comma that separates our weeping and laughing and our breaking down and building up, I have discovered, in my life, we can sometimes treat our commas like they are periods. Because of the many punctuations we may face in our lives, we can sometimes mistake and misplace them. As a result, we remain in some seasons longer than we were meant to be there. We spend too much time living our lives on one side of the comma. We stay in places and relationships God never intended for us to stay in. You have to conquer the commas in your life. God is trying to get you to raise your head and lift your eyes to see that there is so much more to your life than what you are accepting right now. Stop misplacing your punctuations because this is not the end.

> **NR**
> We spend too much time living our lives on one side of the comma.

Imagine this. God has written a story, a masterpiece, for our lives. Each of us plays a special role in this unique story. For most of us, this story is a novel and for a few others it's more like a drama. As we travel through this story, we, many times, arrive at junctures that force us to make difficult decisions. We arrive at places in our story that seem like a dead end, but in actuality, it is just the closing of the chapter. For many of us, transitioning to the next chapter in our lives is and has been the most difficult thing that we have had to do. Having to let go. Having to trust God. Having to move on.

For others, moving to the next chapter has been great. It was expected. You feel excited. But in this journey of uncertainty, just like any great story, there is a rising action, a climax, and a falling action. You will go through valleys in your life; you're going to climb some mountaintops. You are going to meet people that in your heart you hoped would be around forever. But as the seasons change, like leaves, they have to fall away. You are going to experience hurt, pain, and disappointment in this life. You must remember where there is a comma, do not place a period. God is still speaking. There are people we hope will always be part of our lives, and people we feel we cannot let go of. But where there is a period, do not change the punctuation. God wants you to move on because there is something better waiting ahead of you.

TEENAGE MUTANT NINJA RELATIONSHIPS

I'm reminded of my favorite childhood cartoon, "Teenage Mutant Ninja Turtles." When I first came to Memphis, Tennessee as a little boy, my best friend Corey Richmond had all of the Ninja Turtle accessories and action figures. Anything related to the Ninja Turtles, Corey owned. I mean, he had so many of the turtle toys that I would go home and instantly have an attitude with my mom because I felt like she didn't love me like Corey's mom loved him. I had only one turtle, Leonardo, and he wasn't even my favorite. I was always jealous of Corey but not so jealous that I refused to play with his toys. We lived for Saturday mornings. When the sun would rise I would run over to his house and knock on the door and ask Corey to come outside to play. I brought my one little Leonardo turtle while he brought out all of his. We would play all day long.

If you don't know anything about the Teenage Mutant Ninja Turtles, here is a brief synopsis. There are four turtles located in the canals of New York City. One day they are exposed to a mysterious mutagenic gel called ooze. When they were exposed to this ooze, they were transformed into these giant, strong, fighting turtles. After encountering this mutagenic, toxic gel, they mutated into large turtles that could walk upright, talk, and eat pizza. They became furious with evil and lived to fight crime. They were mutant ninja turtles and I wanted to be just like them. I wanted to be bigger, stronger, walk upright, fight crime, and eat pizza all day long. When I think of being exposed to toxic relationships and toxic people, it's just like becoming a Teenage Mutant Ninja Turtle.

Actually, they're nothing alike.

There are no similarities between mutant ninja turtles and toxic relationships. The reason toxic people and relationships are not like mutant ninja turtles is because when you are exposed to toxic relationships and toxic people, you don't become bigger, stronger or walk upright. You don't learn how to fight, or gain superhero strength. You don't become immortal. Unlike the ooze, when you are exposed to toxic people, it has the opposite effect. You become weaker, you become fatigued, happiness flees from you, and instead of walking upright you are likely bent over and emotionally spent. When we allow ourselves to become closely acquainted with toxic relationships, we lose life, we don't grow, we can no longer maximize our talents, and our identity slowly dwindles away. The only similarities between the mutant turtles and our mutant relationships, to my discovery, is that you will find yourselves living in dark places and only associating with other people who live in dark places.

Toxic is defined as anything acting as or having the effect of poison. Some poisons, or toxins are fast-acting, while other poisons can have a slower effect. Acute toxicity, or poison, is when you encounter poison once or twice in a short span of time. It has immediate and short-lived symptoms and adverse effects. For example, if a person accidentally inhales or ingests a drug or chemical, they have experienced acute poisoning and should receive immediate medical attention. Chronic toxicity, or poison, is when you have continued or repeated exposure to toxins or poison, and it has long-term adverse effects. One example is a person who suffers from drug abuse. Because they continue to ingest and subject themselves to harmful drugs, they will, as a result, have long-term side effects, or perhaps, experience death.

There are only two consequences of overexposure to toxins: impairment or death. The same goes for toxic relationships. Many of us have suffered from acute toxic relationships - we've only needed to be exposed to one or two toxic people in our lifetime. While there are those of us who suffer from chronic toxic relationships. We have the habit of repeatedly and continuously exposing ourselves to poisonous people and relationships. Just like poison, there are only two results to overexposure to toxic people and relationships - impairment or death. When you allow yourself to remain in toxic situations, they will either cripple and emotionally damage you so that you are unable to trust and function properly in relationships. Or they will kill you spiritually, emotionally, socially, and/or psychologically. Toxic relationships will leave you useless and unable to sustain and perform in a normal relationship because of the dangerous toxins to which you were exposed.

I have a Kenyan friend who recently wrote me. She broke up with her boyfriend and baby's father and the past couple of years he has been trying to get her back. She recently decided that she would return to try and make it work. "But nothing has changed," she says. "He drinks excessively and still becomes belligerent." She explained to me that when they argue he always leaves and doesn't come back for one or two days, without any explanations. I asked her why she chose to return to him. She replied, "Because he begged me for another chance and said he couldn't live without me. He said he would change and we would fix himself." She thought forgiving him meant returning to her toxic relationship.

"This has been the worst seven months of my entire life." My physician and I were discussing the state of her newlywed marriage with her longtime boyfriend of eight years. "We are going through a divorce, now," she explained. She was with her baby's father for eight years before they broke up for two whole years. She ultimately made the decision to return to him and to get married "because of image," she said. "My sister was getting married and she was happy so I figured I should get married too," she confessed. "I honestly thought things would get better if we just got married but a day doesn't go by that we don't argue. He is so jealous-hearted that I can't do anything or practically go anywhere. He's so controlling," she said with great frustration. She looked me and in her eyes you could see the pain as she pondered a safe exit strategy from her marriage.

Many times, we allow ourselves to remain in toxic relationships, when they are the very thing destroying us and keeping us negative and unhealthy. Albert Einstein said, "No problem can be solved from the same level of consciousness that created it." Therefore, no toxic relationship can be cured with the same level of consciousness and dysfunction that created it. You need to address your issues, and if there are no real, reasonable improvements, then you need to get out.

The probability that your toxic environment will become what it needs to be while you're there is slim. Leave.

There are five things you need to know about toxic relationships:

1. You cannot thrive in toxic relationships.
2. You should never feel guilty about removing toxic people from your life.
3. There are no degrees of toxicity. There is no such thing as too toxic or not that toxic. Toxic people and situations are what they are, toxic.
4. You should identify, accept, and actively work to change if you yourself are actually toxic.
5. Stop making excuses for remaining in toxic situations.

IT'S NOT YOU, IT'S ME. NO, REALLY, IT'S ME.

I have a confession. My name is Issac DeSean Curry and I am toxic. I am in a period of my life where I have been in quarantine, isolation, and detoxification. I have been harboring, in some small crevice in my life, toxic chemicals within me that have slowly crippled and handicapped the relationships that are within close proximity to me. I used to love the thrill of saving and rescuing and being the hero. It is who I was at my core. It is who I thought I needed to be. Until, that is, my engagement went south because of it. I didn't immediately hit rock bottom because I avoided having to feel the pain. When I did feel the pain, it brought me to the ugly reality that I had a problem. I'm thankful for the grace of God that allowed me to see this. The worst, and yet, the best thing, is to come to the realization that you yourself are toxic in some way. Maybe you are not, but maybe, just maybe you are. What if you are toxic? What if it's not really the other person although he or she really does exhibit obvious toxic behaviors? What if I told you that you are just like a parasite that can't live or thrive unless it has a host to provide it shelter and nourishment. Parasitic people can't become parasites unless there is a host to allow them to be cultivated. Most people, when they read that this subject is going to be about toxic people, situations, and relationships, they immediately begin picturing another person. They silently pray that this person will read this section. They want to get a copy of the words and send them to people so they can see how toxic they are. What if I told you that your diagnosis could be wrong? When looking at your situation, if you apply the wrong diagnosis, it will cause you to produce an incorrect prognosis. The wrong outlook in your context will ultimately lead to you applying the wrong prescription to correct it. What if it's you, not the other person, who is toxic?

David committed the sin of adultery and murder when he slept with Bathsheba, the wife of Uriah, and ultimately had Uriah killed by the sword. After spending money and other resources to cover up his drama, David returned to living his lavish life as king. He became so used to his toxic behavior that he failed to recognize that he was contaminating others. God sent the prophet Nathan to confront David and he shared a riddle with him. In this riddle, Nathan spoke about a poor man who used his last few dollars to purchase one little lamb. The lamb was so precious that it became the family's pet, it ate at his table, and it laid in the bed with this poor man.

Nathan described to David that this poor little man sacrificed much for his family to care for this lamb, only to have the lamb stolen and killed by a rich man who had many livestock of his own. Nathan used these details to tug at the heartstrings of David and to get David to use words of conviction from his own mouth. David was livid with the rich man from the riddle. It was almost too brutal to watch because David was the man responsible for the very thing he was convicting the poor man in the riddle for. "Nathan said to David, *"It's You, You are the problem"* (2 Samuel 12:7). In essence, Nathan wanted David to understand that before you point your finger trying to convict somebody else, be sure that the problem is not with you.

It's not you, it's me. No really, it's me.

Coming to the realization that I am toxic does not feel good. Who wants to hear that they are toxic, poisonous, or unhealthy? Who wants to admit that they are the one who gives room for the toxicity that is present in their lives? In my lifetime, I used to get away with being able to tell a woman, "It's not you, it's me" as a way to escape a relationship. Now, it's not a game, it has truth and relevance to it. You can't keep looking the other way and turning a deaf ear. You can't keep blaming other people for the things you are responsible for. It's not always everybody else. You are not an honorary martyr. This isn't martyrdom. You can't always be the victim.

Could it be that you are responsible for what has taken place in your life? The only thing that your past failed relationships have in common is, you. When you have the wrong diagnosis, you will always have an erroneous prognosis, which will, 100 percent of the time, result in your using the wrong prescription to heal your wound. Think about it. If you have symptoms of the flu: cough, fever, headache, chills, and body aches, it does you no good to go to the local pharmacy to purchase cough medicine. Your symptoms will remain and worsen. However, if you avoid the temptation of impulsively responding to each symptom and try to get to the source of your problem, you will learn a better and more effective way to find healing. You have to change your outlook. If you've been applying the same prescriptions over and over again to fix your issues and they seem to keep recurring in every relationship that you enter, there is nothing wrong with the medication. It is just the wrong prescription for your situation. Start with yourself instead of somebody else.

I CAN'T BE WHO YOU WANT ME TO BE

Have you ever remained in a relationship because it was convenient? I mean, you weren't happy or progressing, but you remained there because you felt like you just couldn't do any better? Have you ever been in a relationship that you knew brought out the worst in you? When you looked in the mirror you just did not recognize the person you were looking at? Have you ever been in a relationship where you just felt stuck? You had no ceiling to grow? You were not moving anywhere? It seemed like life was just passing you by? Have you ever wanted to walk away from a relationship but you decided to stay primarily out of guilt? You felt like you owed it to the person to stick around? You felt like if you left the relationship they wouldn't be able to survive, as if you are God's gift to a man or woman. Have you ever found yourself bound to a relationship merely because of image? When people look at you do they think you are happy? When you look at your friends and their marriages on social media, do you think that they look happy, so you say, "I'm going to stay and look happy because I want to have the image of happiness."?

Section Three: GETTING PAST YOUR PAST

I remember sitting in a counseling session when Tom looked over at me and said, "I can't find who I was. I'm lost." He was referring to his marriage that was broken and his wife who was divorcing him. "I didn't realize how much I changed with her," He said. I stopped playing the banjo because she said it made too much noise. I stopped my hobby of building wooden toys because she said I spent too much time doing it. I was supposed to sit idle and be more of a homebody. I even became addicted to medication because she convinced me that there was something wrong with me." Have

> **NR**
> The relationship that makes you feel you aren't enough and that you must become something you're not in order for there to be happiness, is an unhealthy relationship.

you ever been in a relationship where you have found yourself jumping through hoops and performing acrobatic stunts and circus acts just to be affirmed and received by the other person? Have you ever been in a relationship where you felt like "being you" just wasn't enough?

One of the most dangerous relationships is the one that forces you to assume an identity that doesn't belong to you. The relationship that makes you feel you aren't enough and that you must become something you're not in order for there to be happiness, is an unhealthy relationship. One of the Bible's best kept secrets about unhealthy relationships is the one that sheds light on Rebekah and her son Jacob.

Genesis 27:5-27, NLT

5 But Rebekah overheard what Isaac had said to his son Esau. So when Esau left to hunt for the wild game, 6 she said to her son Jacob, "Listen. I overheard your father say to Esau, 7 'Bring me some wild game and prepare me a delicious meal. Then I will bless you in the LORD's presence before I die.' 8 Now, my son, listen to me. Do exactly as I tell you. 9 Go out to the flocks, and bring me two fine young goats. I'll use them to prepare your father's favorite dish. 10 Then take the food to your father so he can eat it and bless you before he dies."

11 "But look," Jacob replied to Rebekah, "my brother, Esau, is a hairy man, and my skin is smooth. 12 What if my father touches me? He'll see that I'm trying to trick him, and then he'll curse me instead of blessing me."
13 But his mother replied, "Then let the curse fall on me, my son! Just do what I tell you. Go out and get the goats for me!" 14 So Jacob went out and got the young goats for his mother. Rebekah took them and prepared a delicious meal, just the way Isaac liked it. 15 Then she took Esau's favorite clothes, which were there in the house, and gave them to her younger son, Jacob. 16 She covered his arms and the smooth part of his neck with the skin of the young goats. 17 Then she gave Jacob the delicious meal, including freshly baked bread. 18 So Jacob took the food to his father. "My father?" he said. "Yes, my son," Isaac answered. "Who are you—Esau or Jacob?" 19 Jacob replied, "It's Esau, your firstborn son. I've done as you told me. Here is the wild game. Now sit up and eat it so you can give me your blessing." 20 Isaac asked, "How did you find it so quickly, my son?" "The LORD your God put it in my path!" Jacob replied.

21 Then Isaac said to Jacob, "Come closer so I can touch you and make sure that you really are Esau." 22 So Jacob went closer to his father, and Isaac touched him. "The voice is Jacob's, but the hands are Esau's," Isaac said. 23 But he did not

recognize Jacob, because Jacob's hands felt hairy just like Esau's. So Isaac prepared to bless Jacob. 24 "But are you really my son Esau?" he asked. "Yes, I am," Jacob replied. 25 Then Isaac said, "Now, my son, bring me the wild game. Let me eat it, and then I will give you my blessing." So Jacob took the food to his father, and Isaac ate it. He also drank the wine that Jacob served him. 26 Then Isaac said to Jacob, "Please come a little closer and kiss me, my son."
27 So Jacob went over and kissed him. And when Isaac caught the smell of his clothes, he was finally convinced, and he blessed his son. He said, "Ah! The smell of my son is like the smell of the outdoors, which the LORD has blessed!

So often, we move into relationships with people and try to become their everything. We wear masks. We alter our plans. We take up new hobbies. We change friends. We wear different clothes. We talk differently. We assume different identities. Because we love and adore the other person, we place unrealistic responsibilities upon ourselves to do things and become people we were never meant to become. In other instances, we enter relationships with people who want to impress and please us so badly that we allow them to do the very thing somebody else may have done to us. We allow them to become somebody they were never meant to be. I've been there before. I've dated women who wanted me to be so pleased and happy with them that they would be willing to, in essence, trade their values, because they thought that would make the relationship flourish. Our culture teaches us that this is the road to happiness. The things we will do just to hear a person say, I love you. The truth is, no matter where you go or who you encounter, you deserve to be with a person who accepts you for who you are - flaws and all. You have to be willing to tell a person, "I can't be who you want me to be."

In the text with Rebekah and Jacob, we observe how easy it is to try and assume an identity that doesn't belong to us just because we desire to please somebody that we love. This narrative has principles that can easily be applied to our romantic relationships. Rebekah insists on Jacob assuming the identity of Esau, because she was secretly disappointed that her and Esau's relationship wasn't unfolding the way she had hoped it would. There was a void that she tried to get Jacob to fill. Jacob, on the other hand, recognizes that he is not his brother. He is keenly aware that Esau looks different, wears different types of clothes, and has a totally different set of skills, but he still tries to become his brother because he simply wants to please the woman in his life, his mother. Jacob was unable to stand up to his mother and tell her, I can't be who you want me to be. He tried to, but eventually he gave in. Because he allowed this unhealthy relationship to flourish in his life, it ultimately robbed him of his peace of mind. Jacob proceeded to allow his mother to dress him and to fix him up so that he could convince Isaac he was somebody he wasn't.

> Our responsibility is to allow people to be themselves and appreciate them for who they are. When we begin to do this, we are making a conscious effort to choose happiness.

How many times in a relationship have you given in, like Jacob, trying to fill the role and responsibility of somebody you weren't? How many times have you tried to fix yourself up and decorate yourself so that you wouldn't disappoint the person you loved? The number one thing that causes pain in any relationship is the expectation we have of other people and who we think they should be, versus who they really are. Once you realize that nobody can be exactly who you want them to be and act exactly how you want them to act, you can begin to alleviate a lot of self-inflicted drama. Our responsibility is to allow people to be themselves and appreciate them for who they are. When we begin to do this, we are making a conscious effort to choose happiness.

Unlike Jacob and Rebekah in the text, you must learn to:

1. Stop dressing yourself up.
If you are spending most of your time changing you clothes, modifying your goals, putting on the right makeup to cover up your flaws so that you will be accepted by the other person, then most likely you are in the wrong relationship.

2. Embrace your true identity.
When Jacob lies to Isaac, he is also lying to himself. As a result, he is forced to flee and run away from the wrath of his brother. When you make a decision to try and be somebody that you are not, you will always find yourself running from your identity. When you try to become somebody else, you will eventually get tired. Stop running from who God has formed you to be.

3. Acknowledge that you can't fix other people.
Rebekah teaches me that, in relationships, I have the power to make a person become somebody that they are not by putting them in a position in my life that they were never meant to occupy. She does this because she is preoccupied trying to fill the gap of unmet expectations left by another person. Who are you trying to dress up in your life and force to be somebody that they clearly are not going to be? Who are

you trying to change and make to fill a void left by another person? Stop trying to fix people.

YOU CAN'T COME WITH ME

There are five words that probably hurt just as much to say them as it does to hear them; "You can't come with me." If you think about it, whenever you hear these series of words in a sentence, it's usually accompanied by some sort of emotion and is not typically associated with the greatest experience. The first time I can remember hearing these piercing words, they came from my father. He said these words as he was leaving me, only for me never to see him again. Although I am accustomed to saying "I haven't ever met my father," I did once before. I've grown accustomed to saying "I haven't ever met him before" because it feels that way. I remember the first and only time I had a chance to meet him. I was in the third grade and attending Sheffield Elementary School. My sister wasn't in school on this particular day, so I remember walking home by myself. When I arrived home, there was a strange man reclining in my mother's chair in her bedroom. My mother then proceeded to introduce me to a man she called, "My Father." I was young and had spent so much time being angry at him because of his absence, but at this very moment, all I wanted to do was embrace my father and never let him go.

I remember his offering to take me to the corner grocery story to buy me some food. I was so excited. We walked and talked. I vaguely remember the conversation but he asked me questions like, "How have you been? How are you doing in school?" I was so excited to be with my father I could barely keep a straight face as we walked. I remember arriving to the store thinking that we was going to lavish me with treats and candy and junk food, but he only offered to buy me a 25 cents bag of Fritos Lays potato chips. I admit, I was disappointed. When you are a child and you finally meet the man who is supposed to be your father, you expect fanfare and all types of gifts to make up for lost time. I guess the only gift he was able to give me was the gift of letting me go. As we began to return home, I could sense he was about to leave because his conversation sounded like he was giving me a benediction. So I politely interrupted him mid-conversation, "Dad, can I go with you?" He paused and acted as if he didn't hear me. He continued to talk and to explain to me how he would be coming around more, but I interrupted him again. "Dad, can I go with you?" After my asking him this question for the third time, he stopped, kneeled down and looked me in my eyes and said, "Son, I'm sorry but you can't come with me." After about a 30-minute encounter with my father, he left and I never saw him again.

I felt a crushing feeling when I heard those words come from his mouth, "You can't come with me." Looking back, I want to believe it was just as hard for him to say those words to me as it was for me to hear them. It's even difficult to say these words to my two year old niece who peers from the window in the storm door every time I prepare to depart from my sister's home. She asks me, "Can I go with you?" Each and every time it breaks my heart to leave her and to have to tell her, "You can't come with me."

It's one thing to have to say these five words to family and to friends. It is a completely different thing when you have to face an important relationship in your life, and have to utter the words, "You can't come with me," with no intention of returning. One of the most difficult things in life is having to let go of a relationship with something or somebody with whom you've invested blood, sweat, tears, time, and energy trying to make work. To have to tell someone who is toxic and who does not bring out the best in you, "You can't come with me," can sometimes be a complicated thing to do. When I, in my past, have conjured the unmitigated strength to speak these five words to someone whose season in my life had eclipsed, I often noticed my actions would sometimes fail to intersect my words. There are times I've tried to let go only to realize that I haven't. Have you ever tried to dismiss somebody but every time you look around you're still holding on to them?

> A second type of toxic relationship is the one you attempt to hold on to long after it has reached its expiration date.

A second type of toxic relationship is the one you attempt to hold on to long after it has reached its expiration date. There are dire ramifications when we try to carry people into our new season when the relationship has already run its course. Another of the Bible's best kept secrets about unhealthy relationships that can help us in our approach to dating and learning how to let go, is the relationship between Abram and his nephew Lot.

In Genesis 12:1, NLT, God instructs Abram, *"Leave your native country, your relatives, and your father's family, and go to the land that I will show you."* In verse 4, the Bible says, *"So Abram departed as the Lord had instructed, and Lot went with him." Abram took his wife, Sarai, all his livestock, and his nephew, Lot.*

I can understand his wife accompanying him on this journey, but tension arises with Abram's need to bring Lot along on the journey with them, too. God previously instructed him to leave his relatives behind. Abram exhibits the reality of the difficulty of knowing someone's time has expired and yet we try to maneuver God's will to meet our expectations.

In chapter 13, after spending years in Egypt, Abram, Sarai, and Lot travel south to live. The topography of their relationship has changed. Abram and Lot find themselves in tumultuous conflict because there is no longer enough room for the two to coexist. (I mean, you can only carry people who have reached their expiration date with you for so long before there will be unceasing conflict).

Some people can't go where God is trying to take you.

The Bible says, *"But the land could no longer support both Abram and Lot with all their flocks and herds living so close together. So disputes broke out between the herdsmen of Abram and Lot." (Genesis 13:6-7).*

Things became so uneasy in their relationship, Abram finally musters up the strength to tell Lot, "This is where I leave you."

Abram's relationship with Lot shows us:

1. Some people can't go where God is trying to take you.
Abraham was trying to take Lot to a place where there was only enough room for him. There was never really enough room for the both of them but neither was willing to face the truth. Circumstances forced Abram and Lot to be honest with themselves about the fact that their time together had reached its expiration date. Just like Abram, we find ourselves trying to carry people with us even after God has shown us

we are supposed to be alone, by ourselves, and going in a different direction. God is trying to take you someplace but where God is trying to take you, everybody can't go with you. That person you've been trying to bring with you can't come with you. You can't bring into your new chapter some of the people and relationships that were meant to stay in your last chapter.

2. Separate yourself so God can reveal His will to you.

In verse 9, Abram says to Lot, *"Please let's separate. Wherever you go, I'm going the other way."* The turning point in the narrative comes at verse 14, when God waits for Lot to depart from Abram before God reveals to him the extent of the blessing He has for him. *"And the Lord said to Abram, after Lot had separated from him, 'Lift your eyes and now and look from the place where you are - northward, southward, eastward, and westward, arise and walk in the land for I give it to you.'"* It wasn't until Abram gained the courage and the fortitude to separate himself from Lot and was willing to go in the opposite direction, that God spoke to Abram. What God is trying to tell you is that he can't show you some things until you first cut all ties from that toxic relationship. Because God covets your undivided attention, you have to completely separate yourself before you can behold some of the things God may be trying to reveal to you. It was as if God knew Abram had been weighed down, distracted, and burdened. "Lift your eyes...arise, walk in the land." Sometimes you can find yourself attached to people that you know you shouldn't be attached to, but it's not until you let them go that you are able to feel some sort of freedom. When you operate in that freedom, God will deal with you and work with you and reveal to you what He has for you. God waited until Abram's distraction was eliminated before He shared His will. I heard someone say this yesterday and although it was figurative in nature, it applies to this moment: " You can't keep the number of your drug supplier and expect to kick your habit." In other words, you can't keep the contact information of the person you know is toxic to you and expect to break away from the relationship or to somehow experience freedom. By keeping their number, you are telling yourself that you don't have the power to break away from that toxic relationship.

3. If you leave the door cracked, you will always go back.

In chapter 14, war breaks out in the region where Lot was living and as a result, Lot was taken captive by the victorious invaders. The text indicates that one of Lot's men escaped and reported everything to Abram, who happened to be living somewhere in relatively close proximity. When Abram learned of this tragedy, *"He mobilized the 318 men who had been born into his household. Then he pursued the army."* Abram was successful in defeating the army and recovering Lot and all of the goods and captives that had been taken. Whenever there is a war, there are always going to be casualties - people who are injured, wounded, and who will consequently, lose their lives. There were hundreds of men who went to war just to bring back Lot at Abram's behest.

If you're not intentional in severing ties, you will find yourself drawn back into that relationship. If you keep returning to that relationship, eventually somebody's going to get hurt. It's the casualty of remaining in toxic relationships. Like Abram, you will end up spending all your time, energy, and resources trying to save and fight somebody's else's battles. If Abram left Lot back in Haran like God asked him to, he wouldn't have gone through the unnecessary trouble he went through.

Why was Abram so attached to Lot that he was willing to ignore the direction of God, carry unnecessary burdens, and fight battles that didn't belong to him? Why are we so inconveniently glued to people we shouldn't be? Could it be that his attachment to Lot was a direct result of his inability to trust God with the details? Could it be that he wanted a child so badly that he was willing to tolerate just about anything just to have somebody fill that void? Both Ishmael and Isaac had not yet arrived to the scene, so perhaps, Abram desperately wanted somebody to call a child of his own. Sometimes our own overwhelming desire and longing for something can cloud our judgment and make us do crazy things just to get them. You can want a spouse so badly. You can want a family so badly that you feel you need to stay in a toxic relationship because you fear it's probably the best that you can do in order to get the family and child that you want. If this is so, then you aren't trusting God with the details. You don't have to put up with average, nonsense, drama, abuse, or neglect just to get what God has for you; don't let anyone tell you otherwise.

After choosing to trust God, in chapter 15, and after Lot is written out of the narrative, God then reveals to Abram that He has a child for him. The blessing of Isaac could not have come until Abram fully trusted and obeyed the voice of God, and until he was willing to tell the necessary people, "you can't come with me."

Section Three: GETTING PAST YOUR PAST

STOP IGNORING THE ALARMS:
5 signs you're in a toxic relationship

A few months ago, I was hosting a meeting at my church with a few friends to plan our next mission trip to Africa. When the meeting concluded, my friend Barry and I began talking as we were preparing to exit the church. We were walking down the stairs and I noticed that the lights from the fire alarm began to flash. Since the alarm had been triggered several times throughout the day, I knew this was a false alarm. I stopped to peek my head into one of the nearby classrooms when a group of people casually looked at me and asked, "Hey, what do we need to do?" I said, "Oh, it's a false alarm, we have been having these all day so don't worry about it." They continued their meeting. When Barry and I continued our conversation, I noticed he looked a bit agitated. "Barry," I asked, "What's wrong?" He said, "The fire alarm is such an important emergency to be diluted by a false alarm." Barry was frustrated because the alarm had been going off and everyone around the entire church was operating business as usual.

Because Barry installs alarms and security systems in his field of work, he understood the mechanics and the details that go into a single alarm being triggered. He explained how the fire detection system is designed to respond to a certain condition of smoke or heat. So when a foreign object or substance enters the inner chamber of a smoke detector, it activates the detector's sensors, which sends a signal to the fire control panel and sets off the alarms. Many of us have experienced false alarms in our lives, so much so, that we don't take the real alarms seriously.

Just like the fire detection system, we all have been given an emergency detection system called common sense. God has given us intellect, emotions, and the ability to know when something isn't right. These alarms are often triggered to tell us that a relationship is toxic, parasitic, and the person isn't really good for us. We know that the lies which are being told to us are not true. We know when the person we are giving our time to doesn't deserve it. Our alarms often go off, and bright red flags are often raised, though we still choose to ignore them. We sometimes treat our real alarms as if they are false alarms, but I am here to tell you, this is not a drill, your relationship may be toxic.

I wish somebody had been there to tell Samson, one of the last judges of the Israelites, to stop ignoring the alarms to his toxic relationships and his toxic behaviors. There were sirens going off all around him but he was so in love that he chose not to heed them. When Samson finally responds, it was too late. Ignoring the alarms to the toxic relationships ultimately cost Samson his life. Ignoring alarms to toxic people can cost you your life. In a brief observation of the life of Samson, I want to identify five principles that can serve as flag posts for you as you strive to become and remain toxic-free.

In Judges 14-16, Samson is enraged. He has rushed into a marriage and has lost a bet to his groomsmen. Samson's wife was forced to share the secrets to his riddle, because of this lost bet. He consequently abandons his wife, goes on a rampage and destroys the property of his enemies. He ultimately falls into a state of depression and erratic behavior. In the span of three chapters, Samson provides us snapshots of three different relationships that utterly fail because they were toxic.

Section Three: GETTING PAST YOUR PAST

1. When verbal weapons are used and physical force is an imminent threat

Judges 14:18 – *"'If you had not plowed with my cow, you would not have solved my riddle.' Then the spirit of the Lord came over him and he went down to Ashkelon, killed thirty men, took their belongings, and gave their clothing to the men who solved their riddle."*

In Samson's moment of anger, he compares his wife to a cow. He refers to her as a heifer. Through time, many words morph and change meaning, but a cow has always been a cow and a heifer has and always will be a heifer. There was a time in the text where Samson referred to his spouse as his wife, but in his anger, he demeans her and calls her out of her name. You're in a toxic relationship when the other person uses verbal weapons such as cursing, name calling, derogatory comments toward you, and there is an imminent threat that he will use physical force. When you are in a relationship where the other person treats you like their personal verbal punching bag, and disrespects you, it's a toxic relationship.

2. When you're afraid to disagree with him

Judges 14:19c – *"Samson was furious about what had happened, and he went back home to live with his father and mother. So his wife was given in marriage to the man who had been Samson's best man at the wedding."*

Judges 15:1 – *"Later on, during the wheat harvest, Samson took a young goat as a present to his wife. He said, 'I'm going into my wife's room to sleep with her,'* but her father wouldn't let him in. 'I truly thought you must hate her' her father explained. 'So I gave her in marriage to your best man.'"

When Samson got upset with his wife, he abandoned her. He was a no call, no show. Without explanation, he departed and didn't even care to tell his in-laws his whereabouts. As a result, life went on without him. When he returned to see his wife, the father stopped him and basically said, "Wait, hold on here. I was sure that when you left in your usual anger and wouldn't talk to her, you were quitting the relationship. When you withdrew and dismissed yourself because you were upset, I figured it was time for my daughter to move on."

> **NR**
> When you're afraid to disagree or to make a mistake in the relationship because every time you do, the other person punishes you by threatening the commitment of the relationship, that relationship is toxic.

When you're afraid to disagree or to make a mistake in the relationship because every time you do, the other person punishes you by threatening the commitment of the relationship, that relationship is toxic. That is an alarm trying to get your attention. It's not a safe relationship if you have to worry about him withdrawing emotionally and even physically every time you don't agree. It's not a safe relationship if they withdraw because you have a competing thought, because you do something he doesn't like, or because you fail to do something he may want you to do. This is emotional blackmail, and Samson was guilty of this when he threatened the commitment of the relationship with his wife, when they had a disagreement. As they had an argument, they didn't talk, they didn't converse, they didn't come to an agreement, Samson left. Obviously he thought it was okay because after time passed and no call, no email, no text, he reappears expecting for things to be normal. He dismissed himself and returned on his own terms.

You are in a toxic relationship when you are afraid to disagree, when you are afraid to have an argument with the other person because you fear you will be punished.

3. When sex is the only way you communicate

Judges 16:1 – *"One day Samson went down to Gaza where he met a prostitute and went to bed with her."*

And so after Samson departed his failed marriage, he had a brief encounter or relationship with a woman of Gaza. The only thing we know for certain is that they slept with one another. There was little to no conversation. The woman doesn't even have a name. We don't know if they had anything in common other than their desire or love for sex. Samson's way of coping with his past heartbreak was to enter into another emotionally detached relationship with somebody else. This relationship was dominated by sex. Another alarm which helps to indicate that you're in a toxic relationship is when the only way you communicate is through sex. Sex is the only time when you both are on the same page, and how you've learned how to cope with the dysfunction in your life. When the only way you know how to handle your problems, both big and small is through sex, you are exhibiting toxic behaviors.

4. When it's always about them and never about you

Judges 16:15-16 – *"Then Delilah pouted, 'How can you tell me, 'I love you,' when you don't share your secrets with me? You've made fun of me three times now, and you still haven't told me what makes you so strong!'" She tormented him with her nagging day after day until he was sick to death of it."*

Samson quickly found himself entangled in another relationship where he willfully ignored the sounds of the sirens wailing for him to get out. In his short-lived romance with Delilah, Samson disregarded her overbearing desire to gain access to his most prized possession, his strength. Her only agenda was to gain wealth by surrendering Samson to the Philistine army. She was only thinking about herself. On three different occasions she nagged him to give her what she desired, and ultimately he did. You're in an unhealthy, toxic relationship when you are connected with somebody who only focuses on themselves and what they have to gain. Are you ever the center of attention? Are your desires and passions ever celebrated? Is the relationship imbalanced? You have to be able to determine whether a person loves you or they just love what you can do for them.

> **NR**
> You are in the middle of a toxic mess when you allow the other person to lie to you.

5. When they lie to you

If they lie once… **Judges 16:18** – *"Delilah realized he had finally told her the truth, so he sent for the Philistine rulers. 'Come back one more time,' she said,* *for he has finally told me his secret.' So the Philistine rulers returned with the money in their hands."*

Delilah lied to Samson on three different occasion before her lie culminated in Samson's death. You are in the middle of a toxic mess when you allow the other person to lie to you. No relationship that is built on lies can survive. If they lied to you once, there is a high probability they will lie to you again. Lying destroys trust and all healthy relationships must be built upon trust and respect.

IS THIS RELATIONSHIP GOD'S WILL?

"How do you know when God is trying to close the door on a relationship," she asked me. Trying to quickly justify and convince herself that the relationship wasn't over, she continued, "Do you think God can forgive me and put me and Paul back together? I know I said he didn't treat me right but can't God change his heart?" As if anything I were to say would make any difference, she followed, "How do I know if this relationship is the will of God?" At this moment, I knew there wasn't anything I could say over the telephone that would convince Charity it was time to put the relationship to rest.

Don't wait until the check engine light comes on before you run a diagnostic on your relationship.

Many times, we attempt to seek God's will for an answer to a relationship inquiry, but we often don't wait around to see if there is a declined letter in the mail. Am I where I should be? Is this relationship for me? For many of us, God may not speak audibly to inform us of this answer, but there are three questions you can ask yourself that will help you to know whether the relationship you are in is indeed a relationship God would endorse. You have to be willing to ask yourself the difficult questions and you need to be able to deal with the answers.

1. Is this relationship a good thing for me?

Don't wait until the check engine light comes on before you run a diagnostic on your relationship. You need a checkup to make sure you're still heading in the right direction. You don't want to be one, like many I counsel, who looks up one day and asks themselves, "How did I get here?" Are you healthy or are you more stressed? Are you inspired to achieve or have you become more complacent? Is your personal growth embraced and celebrated or is it overlooked and demeaned? Do you feel free or more like you're in bondage? Can you be vulnerable in your relationship or do you have a fear to speak your mind? You need to have the courage to know when your relationship isn't right for you.

2. Is this relationship beneficial to me?

A mutualistic relationship is when two organisms "work together," each benefitting from the relationship. An example of a mutualistic relationship is between an oxpecker (type of bird) and a rhinoceros. The Oxpecker lands on the rhino and eats the parasites that nest on their skin. The oxpecker gets food and the rhino get pest control. This example isn't to feed our desires for a "friends with benefits" relationship, but to create a healthy conversation about committed relationships that thrive off of mutual respect and mutual accountability. What do you like about your relationship? What does your relationship do for you? How has investing time and attention to your relationship benefited you? How has it benefited the other person? Are you doing all the work? Is your relationship equally and mutually beneficial to both people?

3. Is this relationship drawing me closer to God?

Is your spiritual growth valued and celebrated? Is your love for God deepening or waning? Are you praying more or praying less? Are you trying to carry the spiritual load for the both of you? If your relationship is not drawing you closer to God, then your relationship is sin for you.

You must trust God to help you accept when a relationship has expired and is no longer in your best interest. You must be able to ask God for the courage and fortitude to release yourself from the relationship and to give you the power to walk away. Although many people may not understand your decision to leave that relationship, seek God to help you to remain steadfast and to keep from turning around and returning to that relationship. I like the way Melody Beattie says, *"it takes courage to end a relationship - with friends, loved ones, or a work relationship. Sometimes, it may appear easier to let the relationship die from lack of attention rather than risk ending it. Sometimes it may appear easier to let the other person take responsibility for ending the relationship."* Have the courage to end relationships that are not God's will for you.

LOYALTY IS NOT LOVE.

Proverbs 26:11 - Just as a dog returns to its vomit so a fool repeats his foolishness.

Martyr - mahr / ter - 1. a person who undergoes severe or constant suffering. 2. a person who willingly endures great suffering or death rather than renounce his or her belief, principle, cause, or religion.

A Romantic Martyr is what I consider a person who, for the sake of maintaining a relationship and a false sense of security, would rather tolerate neglect, isolation, abuse, and/or pain than renounce the relationship. I remember seeing my mother kneeling over in the chair clinging to the banister as I raced up the stairs. Her dress was torn, her hair in disarray, whips adorned her back, and her eyes swollen because tears perpetuated them through the previous night. The only thing I could do was embrace her and help her into the house. After having her boyfriend drag her out of a nearby restaurant and physically abuse her, she spent the entire night locked in his apartment while he was at work. I consider my mother one of the strongest women I have ever met, but even she returned to him.

> **NR**
> Your ability to endure and remain in your dysfunctional relationship is not a reflection of love but is a dimly lit silhouette of your loyalty. Love and loyalty are not one in the same.

I have been trying to piece together a better portrait of romance and love because our culture has a screwed up ideology of what a healthy relationship looks like. Somehow we've learned to become Romantic Martyrs. We've developed the belief that relationships have to be hard and difficult in order to be successful. This is the mindset of a martyr. Somehow we think the more difficult our relationship, the more credit we deserve for remaining loyal. We create dysfunctional competitions among ourselves, sort of like a competition among martyrs.

To put a relationship before your own values, beliefs, and needs is borderline idolatry. To overstay and remain in a place that is not healthy for you and to elevate a relationship before and above your own self is worship. Your ability to endure and remain in your dysfunctional relationship is not a reflection of love but is a dimly lit silhouette of your loyalty. Love and loyalty are not one and the same. Why do we confuse love with our need to be loyal to a relationship that is toxic? They are related and yet they are drastically different. Their paths do intersect but they stand alone. They have similar characteristics and share a similar hue but they still are not the same.

Love is reimbursed and shared. Both love and loyalty accompany one another in any thriving relationship, but when they don't, you as the significant other have an important decision to make. Both loyalty and love begin from the same starting position, but love has the ability to stop running because it realizes "we both deserve to be happy." At the same time, loyalty will continue to run blindly because it refuses to accept that "I can do better." Love has the awareness to stop while loyalty runs aimlessly in pursuit of approval, affection, and acceptance. When love and loyalty are no longer partners in the same race, you must decide how much longer you're going to stay. Loyalty is not love. You should never misconstrue one for the other or consider sacrificing one for the other.

NEVER RETURN TO EGYPT

"For the Lord has told you, you must never return to Egypt."
Deuteronomy 17:16c

His name was Robert. He was nearly 20 years my mother's elder. She was crazy about him. In my estimation, she was so crazy about him that she lost herself in him. I remember her damaging her relationship with my grandmother; disregarding and avoiding the advice from family and friends. She surrendered her stability by taking my sister and I and moving in with him. She relinquished all control and power and gave it to him. She gave him the keys to her heart and as a result he took those keys and drove her crazy. Because he had all of the control whenever they would argue or had conflict, our livelihood was at risk. I remember when he put her out. Because she sacrificed all her other relationships for her relationship with Robert, we had nowhere to go. For three months we lived in a motel on Lamar Avenue and we ate cold cuts every day all day. Although we experienced that grief, she still wasn't completely finished with that relationship.

And then there was Anthony. My mother was shaken by him and she does not scare easily. In fact, it had gotten so bad that she couldn't wear the clothes that she had wanted to wear. He would practically dress her every day. She couldn't spend time with family. She couldn't spend time with friends. She pretty much had to be alone. It was so bad, that when he would come over to the house, she would leave us and leave whoever was there and she would go into the room and he would lock her into the room and there they would remain until he decided it was time to leave. Anthony mentally abused and controlled my mother. He changed her spending habits. We all were walking on eggshells around him. My sister and I knew our mother wasn't the same when she was in a relationship with Anthony. We desperately needed her to get away from that relationship because my mother just wasn't herself.

Next there was Jerry; he drove a yellow Cadillac. He was physically, emotionally, and psychologically abusive toward my mother. I remember vividly. I woke up one morning and I noticed that my mom didn't come home from the previous night. She had just ended her volatile relationship with Jerry. And to exhibit her newfound freedom, she decided she was going to go out with her sister-in-law, Marilyn, to a bar and grill located somewhere on Elvis Presley Blvd. I woke up that next morning to a phone call from my mother, she was crying, her voice was muffled, "Sean, Help me, I'm scared." And nothing is more terrifying and heart-wrenching than to hear your mother crying. She said, "I am locked in this apartment, and I can't get out." Then she hung up abruptly. Without thinking or developing a plan of action, I jolted from the apartment and ran downstairs looking for a car to jump into because I needed to find my mom, but I had no idea where Jerry lived. When I couldn't find my mother, I returned home. We lived in an upstairs apartment. So when I arrived home, I ran upstairs and I remember seeing my mother kneeling over in the chair clinging to the banister. Her dress was torn, her body half exposed, her hair in disarray, and whips adorned her back because Jerry took a cord and whipped her. After he dragged her out of a nearby restaurant and physically abused her, she spent the entire night locked in his apartment while he was at work. The only thing I could do besides being enraged was embrace her and help her to walk into the house. I consider my mother one of the strongest women I have ever met, but even she returned to him.

And, finally, there was another Jerry. This Jerry drove a blue car. I don't remember what kind of car it was; I just remember it was blue. This was the longest dysfunctional relationship that I can ever remember my mother being in. They would argue, they would fight, they would break up and then they would make up. To make matters worse, alcohol accelerated everything because he would drink just as much as she would drink. They would fight, he would move out. They would make amends and he would move back in. My mom depended on him, financially, so she tolerated him and had no control over the relationship. I was in graduate school at Princeton during this time so it became very difficult for me to manage my workload and to facilitate a dysfunctional relationship 1,000 miles away from home. This relationship was so draining it drove a wedge between my mother and me. However, one day this guy was bullying my baby sister, and everybody knows, do not mess with my baby sister. She called me in a frantic scare as a last resort, which sent me in a rage. I drove frantically in my vehicle with only one mission: to destroy this man. I could hear nothing. I saw nothing. I felt nothing. When I arrived to my mother's home, Jerry and I fought, briefly, before my sister touched me gently, in a way only she can, in order to bring me back to reality. Through all of this, my mother returned to this relationship.

The most difficult thing for my siblings and me to do was to embrace the fact that my mom was not finished; she wasn't ready to

> **NR**
> To submit yourself to hurt, pain, neglect, isolation, fear, silence, and stagnation, when you know you are worth so much more, is not only settling for less, but is selling yourself for less than your true value.

give up that dysfunctional relationship. Sometimes the hardest thing for people to do is to remove themselves from other people who are in toxic, dysfunctional, and unhealthy relationships because if you remain in close proximity, that stuff splashes onto you. Before you know it, you are repeating some of these same behaviors you see with other people close to you. No matter what my mother did, I chose to fully remove myself because her toxic relationship was making me toxic.

The ultimate betrayal is not when other people betray you but when you learn to betray yourself. When you remain or return to a relationship, knowing you need and deserve better, that is the ultimate self-betrayal. To submit yourself to hurt, pain, neglect, isolation, fear, silence, and stagnation, when you know you are worth so much more, is not only settling for less, but is selling yourself for less than your true value. You are betraying yourself and who God has called you to be. Dysfunctional relationships are equivalent to the Israelites wanting to return to Egypt in the book of Deuteronomy. They knew what it was like to experience freedom, but they didn't know what it was like to remain free. They were so familiar with oppression and accustomed to living in fear, they failed to embrace their freedom. My mom had a knack for being able to pick and select the wrong candidate with whom she would enter into a committed relationship. She left bondage with one person and somehow found herself in bondage with another person.

Nothing is more debilitating than to leave bondage and dysfunction only to re-enter into bondage and dysfunction elsewhere. Nothing is more discouraging than to give your heart to a person and have them to break it, and after piecing together your life again, you give your heart to another person only to have it shattered, almost irreparably. Nothing is more destructive than to have the opportunity to start fresh and anew only to repeat those behaviors, to use the same strategies, and to keep the same model for selecting relationships that are short-lived.

Deuteronomy 17: 14-20 (MSG) reads,
"14 When you enter the land that God, your God, is giving you and take it over and settle down, and then say, "I'm going to get me a king, a king like all the nations around me," 15. Make sure you get yourself a king whom God, your God, chooses. Choose your king from among your kinsmen; don't take a foreigner—only a kinsman. 16 "The king must not build up a large stable of horses for himself or send his people to Egypt to buy horses, for the LORD has told you, 'You must never return to Egypt.' 17 The king must not take many wives for himself, because they will turn his heart away from the LORD. And he must not accumulate large amounts of wealth in silver and gold for himself. 18 "When he sits on the throne as king, he must copy for himself this body of instruction on a scroll in the presence of the Levitical priests. 19 He must always keep that copy with him and read it daily as long as he lives. That way he will learn to fear the LORD his God by obeying all the terms of these instructions and decrees. 20 This regular reading will prevent him from becoming proud and acting as if he is above his fellow citizens. It will also prevent him from turning away from these commands in the smallest way. And it will ensure that he and his descendants will reign for many generations in Israel."

Before arriving at chapter 17, the children of Israel had been captive, oppressed, and entangled in a dead-end, committed relationship with Pharaoh. I use the term, "committed" because I'm trying to draw a picture here. In essence, a committed relationship truly describes the Israelites and Pharaoh. It may have been an unwanted, forced relationship but it was binding to say the least. Moses, by the power of

God, was able to break them free from their "toxic" relationship with Pharaoh. Now after crossing the Red Sea and spending an extended amount of time in the wilderness, they are just about to enter the Promised Land that God destined for them. However, in order for the Israelites to maximize their experience, learn how to prosper, and take advantage of their new season, God wanted Moses to remind them of the advice God had given to Moses, on their behalf, when Moses spent 40 days alone with God on Mt. Sinai (Exodus 24:18). God had given Moses some pointers on how to keep from returning to Egypt, be it geographically, physically, emotionally spiritually, mentally, or psychologically. There are six things Moses wanted Israel to know about their time in Egypt and their possibilities of happiness in Canaan, that we can apply to our relationships, today.

1. A wilderness attitude won't work at a Promised Land address. The children had just come from out of Egypt and through the wilderness. They were about to enter the Canaan land, a new place which would require a new way of doing things - creating a new culture. If they were going to prosper in this new place they could not bring the same attitude and the same behavior that kept them in Egypt and that forced them to wander in circles in the wilderness for forty years. Egypt represents, for us, those relationships or that person that kept us in bondage, and if we are ever going to be able to experience the life that Canaan has to offer us, we must change our attitude and the behavior. You can't flourish in a new relationship still operating with an old attitude. You can't continue to offer the same old disposition and the same insatiable appetite to always be correct and expect for your new relationship to work.

2. You can't allow your misfortunes in Egypt to dictate how you live your life in Canaan.

Moses wanted them to know that they can't allow the abuse, the pain, and the heartache in Egypt to impact their ability to enjoy life in Canaan. If you are going to embrace the things God has for you, you must learn how to live your current life. Whether you have experienced divorce or you were the cause of a terrible breakup, you must learn to release your past so that you can live for the future. Don't allow what happened to you in your past to dictate and control how you live your life today.

3. Making a commitment prematurely will only replicate your Egypt experiences.

God knew that the people were not ready for another committed relationship. They were still experiencing reverberations from their toxic relationship with Pharaoh. Moses wanted them to know that if they rushed into another relationship with a King, they would simply produce another Pharaoh. Moving into another relationship when you know very well you don't have the time or emotional capacity to bear it, all you're going to do is see images of your past relationships. We can selfishly compromise potentially good people and relationships because we irresponsibly pursue them when we should be in our season of waiting and recovering from our last relationship. Whenever you try to enter into another relationship without having properly grieved the passing of another relationship, you are bound to repeat the experiences from Egypt.

4. Just because you find a king doesn't mean you're going to be happy.

God knew that the people would eventually mix their priorities. He knew that they would eventually believe if they could select a king to lead them like other countries, they would, as a result, find happiness like everybody else. This was erroneous and Moses wanted them to know that God should be the source of their happiness. As a business marketing major in college, I learned the art of consumer behavior. Most of the marketing and advertising in the industries of makeup and beauty, health and lifestyle, and romance and family, try to bombard us with unrealistic story plots, captivating colors, and nostalgic music to get us to buy into a 30-second commercial that reinforces the idea that happiness only comes if you do "this, that, and the other thing." We are told that if we use this type of body soap we will attract a certain type of woman, which will result in our ultimate happiness. And we buy into it. We are convinced that we need faster cars, bigger houses, and better paying jobs in order to experience happiness, when we should be able to arrive at happiness and fulfillment aside from these things. Just because you may find a man or a woman, doesn't mean you are going to be happy. You may even get a really good one but you still may never truly arrive at the intersection of happiness. Why? Because God is supposed to be the source of your happiness. If God isn't the source of your happiness, you can find a man, woman, king or queen, and still won't be truly happy. No man or woman can provide for you the fulfillment that God can give you.

> **NR**
> If God isn't the source of your happiness, you can find a man, woman, king or queen, and still won't be truly happy. No man or woman can provide for you the fulfillment that God can give you.

5. Whenever your desire to have a king outweighs your desire to embrace the love of God, you have lost your way.

The children of Israel were overlooking the blessings that God had provided them. He brought them out of Egypt into a new season and gave them everything they wanted and needed, and yet, they wanted more. Whenever your desire to have a relationship or romance outweighs your desire to embrace the love of God or interferes with the love you have for God, you, my friend, have lost your way. Nobody will admit that their desire to be in a relationship outweighs their desire for fellowship with God; however, our actions show it. When you put more time into complaining about your season of singleness than you do about embracing God and God's purpose for your life, you easily overlook what God is trying to do in you today.

A PARADIGM SHIFT

It is important to understand that no king could ever become king unless the people first chose him to be their king. In a similar way, you can't enter into a committed relationship with someone unless you first choose to be in that relationship. In Deuteronomy 17, God was not encouraging Israel to appoint another king to rule their nation; He was actually against the idea. God wanted to be their singular focus. He wanted to be their king. However, God knew that the people would eventually demand a different type of king for selfish reasons. Since they were going to insist on selecting another king, God wanted them to at least choose the best possible king for themselves. Therefore, in order to choose, select, or appoint a king, Israel would be electing to enter into a committed relationship with somebody else. Since selecting a king represents our ability to choose a committed relationship to enter into with somebody, God provides for us critical advice on how to choose and create healthy committed relationships.

God's Model For A Healthy Committed Relationship:

1. Check your motive for wanting to date and be in a relationship. (v.14)

The very first thing God says is, *"When you enter into the land that God, your God has given you, you are going to take it over, then you are going to say, 'I'm going to get me a king like all the people around me.'"* God knew there would come a time when the people's desire for a king would be influenced by societal terms. They wanted to be like everybody else. This culminates in I Samuel 8:5, where the people cried to Samuel and asked for a king that would fill their voids by fighting their battles instead of learning how to trust in God for themselves. Their desires and motives were selfish, misguided, and motivated by impatience. God's model for a healthy, committed relationship begins with first checking your motive for seeking to be in a relationship. If your desire to be in a relationship is so that you don't have to sleep alone, or so that someone can fill your voids and fight your battles, then you want to be in the relationship for the wrong reasons. If your desire for romance stems from the fact that you secretly want to be like everybody else around you who perhaps is engaged or married, you are sadly misguided. When you build your house based upon someone else's blueprint; you will always be unfulfilled and unhappy. When you try to develop and build your

When you try to develop and build your life to look like someone else's life, it's never going to look the same, therefore you will always be left with a void on the inside.

life to look like someone else's life, it's never going to look the same, therefore you will always be left with a void on the inside. You can't build your life based upon somebody else's Facebook photos because regardless of what they look like or what they appear to tell you, you will never know the complete story behind them. When you are building and constructing your life, make sure it's the blueprint that God has for you and not somebody else's.

2. Make sure you select someone God chooses for you. (v. 15a)

"Make sure you get yourself a king whom God, your God, chooses." God had one request that would ultimately influence everything else about the king the people would choose; allow God to select the king for them. The second model for building a healthy, committed relationship God's way is making sure that, through much prayer and supplication, God is the director of your selection process. There are a lot of people to choose from, but my selection process has proven to be severely flawed. I need God's guidance. You want to make sure that the person you're dating is not just somebody you like or who has satisfied the prerequisites on your personal checklist, but that they are somebody God has placed in your life for you. Has God chosen for you the person you are pursuing now or are they just a result of your own goodwill hunting?

You must be selective, restrictive, and most of all, patient, in selecting to be in a committed relationship. You cannot rush a healthy relationship.

3. They have to be of like mind, spirit, and walking in the same direction as you. (v.15b)

"Choose your king from among your kinsmen; don't take a foreigner—only a kinsman." One of the essential rules for Israel in selecting a king was that he came from among them and was not of a different belief system. Although Israel would eventually disregard this point and elect kings outside of God's will, they would learn the significance of choosing somebody that not only looks good but somebody who is good for them. The third model for building a healthy committed relationship God's way is one of the essentials in discovering whether the person is God's choice for you. Make sure you share common interest, common values, and a common purpose. It's not good enough to find a man or woman that simply attends church; that can be superficial. You must take your time to investigate and properly vet a person before you give them your hand in commitment. The purpose of dating is so that you can take your time to retrieve and assess the "data" before you make a serious commitment. You must be selective, restrictive, and most of all, patient, in selecting to be in a committed relationship. You cannot rush a healthy relationship.

4. Pay attention to whether he truly depends on God. (v.16a)

"The king must not build up a large stable of horses for himself." This verse alluded to a King's military power. If he wasn't able to build large stables of horses, this would limit his ability to build a large army. Therefore, if the king wasn't able to depend upon his military power, he wasn't able to entangle himself in political alliances, which meant he would have to be in constant reliance on God. The fourth model for building a healthy committed relationship God's way is paying attention to whether he

truly depends on God. Is his dependence on God a performance or is it personal? In your dating, you need to be able to assess whether or not the other person truly depends on God. You will only discover this if you are patient and if you pay attention to how they handle situations when they are under pressure. This is important because some people will act as if they practice walking by faith, but when times get hard in your relationship, while you're trying to seek guidance, they will be trying to fight the battle with their own hands. This makes a difference. Anybody can trust in their resume and in their skill, but if you're going to give your hand in commitment to anybody you want to know that aside from any and everything else, they seek God in all they do.

5. Pay attention to how he treats the people around him. (v.16b)
"or send his people to Egypt to buy horses, for the LORD has told you, 'You must never return to Egypt.'" Egypt was a dangerous place. People who traveled there very rarely came back alive. For this reason, the king was not only forbidden to build stables of horses for himself, he was prohibited from ever sending any of his people back to Egypt. The fifth model for building a healthy committed relationship God's way is paying attention to how they treat the people around them. There is some validity to the old saying, "If you want to know how a man will treat you, pay attention to how he treats the other women in his life, such as his mother and his sisters." You can learn a lot about a person by observing how they treat the people in their life. If a man treats the woman he is dating with respect but he treats other people around him badly, this should be a red flag. If you love somebody you won't subject them to Egypt under any circumstances. However, if you have the love of God in you, you wouldn't want to subject anybody to Egypt. If you are going to have a healthy relationship, it

is important to make sure that your partner has a healthy love for all people, not just you.

6. Pay attention to what type of company he keeps. (v.17a)

"The king must not take many wives for himself, because they will turn his heart away from the LORD." God highly discouraged the king from having many wives because this would cause him to become distracted and to turn from the will of God. It was paramount that the king kept positive influences around him to keep focused because God desired for the king to be a faithful king. The sixth model for building a healthy committed relationship God's way is paying attention to the type of company the other person keeps. Not only can you learn a lot about a person by how they treat other people close to them, but you can learn a lot about a person by the people they spend most of their time around. If you are going to create healthy committed relationships you must pay attention to the company he keeps and also demand and exhibit faithfulness.

7. Pay attention to how they view and handle money. (17b)

"And he must not accumulate large amounts of wealth in silver and gold for himself." God forbade the king to be a lover of money because he would divide the king's loyalty to the people and to God. The seventh model for building a healthy committed relationship God's way is paying attention to how they handle money. One of the leading reasons people file for divorce in America is because two people have different viewpoints and ways of dealing with money. One is a saver while the other is a spender. One is frugal while the other is more prodigal. One has school loans piled to the ceiling and is fine with it while the other

has no school loans and resents the other. Take your time and pay attention to how he views moneys. How does she handle money? If you find yourself connected to a lover of money, eventually the money will substitute and become their god. If the money can become their god, it won't be long before it will replace you.

8. Make sure they have their own personal relationship with God. (vv. 18-19)

"When he sits on the throne as king, he must copy for himself this body of instruction on a scroll in the presence of the Levitical priests. 19 He must always keep that copy with him and read it daily as long as he lives. That way he will learn to fear the LORD his God by obeying all the terms of these instructions and decrees. 20 This regular reading will prevent him from becoming proud and acting as if he is above his fellow citizens. It will also prevent him from turning away from these commands in the smallest way. And it will ensure that he and his descendants will reign for many generations in Israel." God warned the people to choose a king that had his own copy of the Word of God. Not only should the king have his own copy but he must have it beside him and read it daily. The final and most important model for building a healthy committed relationship God's way is, making sure they have their own personal relationship with God. You're desperately trying to get him to come to church? You've been working hard to get her to pray with you more? You're really hoping and praying that he grows more spiritually? These are all admirable actions, but the only problem with this scenario is that when you began dating, if you had paid attention to the position of his Bible you would have learned more about his spiritual walk then. If you're going to have a healthy and committed relationship, you want to make sure that person has their own personal relationship with God

and that you're not trying to create that relationship for them. You can learn everything you need to know about a person's spiritual walk by the position and condition of their Bible.

CHAPTER 10

HAVE THE FUNERAL

LEARNING TO LET GO AND MOVE BEYOND YOUR PAST

If there is anything relationships have taught me, it's that when dealing with a breakup and experiencing a broken heart, you must be careful because homeless emotions will look anywhere to find shelter. Recovering from my broken engagement had proven itself to be a much more difficult process than I imagined it would be. Months after the engagement was over, I became emotionally attached to another woman. It was never my intention to move in haste, but it happened nonetheless. I found a nice place for my homeless emotions to nest and to rest. I thought I had moved on but I didn't know I was still carrying, toting, tugging and pulling the corpse of my decayed relationship along with me. I became an emotional pallbearer because I was carrying this proverbial coffin from one relationship to another, and inside this coffin was the corpse of my failed relationship. I thought I had let go but really I had hid it away and was carrying it on my back like a tote bag while in another relationship. I was cheating on my future because I was still holding hands with my past.

> NR
> If there is anything relationships have taught me, it's that when dealing with a breakup and experiencing a broken heart, you must be careful because homeless emotions will look anywhere to find shelter.

I had been holding on to a corpse that had been growing older and older by the day. The truth is, I refused to have the funeral for my past relationship. Why? That's a good question. I had not proceeded with any plans of a funeral because I was unable to look at my past and deal with my loss responsibly. I mean, I dealt with it long enough to hide it away. I coached myself to ignore the putridity of deadness growing inside of me. I knew I needed to let go, but I secretly had a hard time getting myself to accept it. Our relationship was so promising. It was, but it was no more. I tried to delete photos of our time in Africa, where we met on missions, but I couldn't. Instead, I tried to bury her artifacts in some lost storage space because I couldn't gain the wherewithal to throw them away. In the following three years, I had seen two more relationships and yet I failed to have one funeral. I was so busy trying to find somebody to shelter my wandering heart and my homeless emotions that I failed to give the necessary attention to dealing with my past.

Many times when we feel betrayed or when we experience the death of a relationship, we become so busy trying to find something or somebody to shelter our hearts that we fail to deal with the damage and the casualties that resulted from the warfare we recently engaged. Through trial and error, I learned:

1. You have to bury the relationships in your life that have died and allow them to rest in peace.
2. You have to stop breathing oxygen into dead places. You have to cut off the oxygen to the relationships in your life that should not still be alive.
3. You cannot recycle and reuse worn emotions you gave somebody else.

4. You cannot resurrect and replicate the heart you gave to your last relationship and try to share it with the person you are attempting to love, today.

5. The people in your present don't deserve a damaged you, they deserve the best possible you.

EXIT STAGE LEFT

While they were eating, he said, "I tell you the truth, one of you will betray me." Greatly distressed, each one asked in turn, "Am I the one, Lord?" He replied, "One of you who has just eaten from this bowl with me will betray me. For the Son of Man must die, as the Scriptures declared long ago. But how terrible it will be for the one who betrays him. It would be far better for that man if he had never been born!" Judas, the one who would betray him, also asked, "Rabbi, am I the one?" And Jesus replied, "Yes, you are the one."
(Matthew 26:21-25, NLT)

One of the best things Jesus ever taught me about relationships was the example of how to press the stop and eject buttons and how to dismiss those people who have served their purpose in my life. He does so without harboring anger or resentment towards others. In a time when he was most vulnerable, Jesus was betrayed by one of the people closest to him, Judas. For years they spent time together, growing in their friendship. They built a relationship with one another's family members and they created memories with one another. In Jesus' awareness that their relationship had reached its expiration date, he immediately released Judas from his presence. Jesus pointed and Judas exited, stage left. Betrayal hurts. When the person you've loved, grown with, and invested in has to leave, it isn't always easy to navigate life afterwards. Jesus shows us how to handle betrayal based on how he handled Judas:

1. Accept that some people are not meant to remain with you.
Jesus didn't get angry, he didn't fight, nor did he lose sleep. He recognized that there was a bigger picture and sometimes the people in your life are not a part of that portrait. Until you open your clasp, you won't be positioned to grasp what God is trying to give you.

2. Allow those people who don't want to be in your life to leave.
Jesus could have attempted to hold onto Judas but what good would that have done? Judas already had his mind made up that he didn't want to be there. The worst thing you can do is try and hold somebody hostage who doesn't want to be in a relationship with you. A person's actions communicate their intentions and desire to be in your life. He teaches us that not everybody is meant to be in our lives forever and when their time has expired, our best course of action is to allow that person to leave.

> **NR**
> A person's actions communicate their intentions and desire to be in your life.

3. Believe that God has a greater purpose and plan for your life.
"For the Son of Man must die." In other words, Jesus was saying, "The pain that I am experiencing at this moment ultimately has purpose in it." When Jesus saw the desperation of the disciples, he tried to encourage them by letting them know that he needed to be betrayed in order for God's will to come to pass. Judas' betrayal led to Jesus'

death on the cross, which provided salvation for humankind. If Jesus had not released Judas, the possibility for salvation could have been thwarted. You can't avoid Judas in your life, because if you avoid the Judas, you avoid God's plan. We can't avoid some of the pain we've encountered because there are some things that God will do through our lives as a result of it. We must believe God has an ultimate plan when it comes to the different seasons of our lives and the people that will help us to get through them. Although Jesus couldn't see God's plan fully, He accepted it. Perhaps you cannot see what God is trying to do through your brokenness but you must accept, allow, and believe that God's plan is greater than what you can see.

Jesus' ability to release people as their seasons eclipsed is something I had desired to do for a long time. It's not easy to accept that a relationship has run its course and to have to release a person. When a relationship is over, we will find any reason possible to hold onto hope and the possibilities of a resurrection. If you are going to be able to successfully have a funeral of your past failed relationship, you must learn how to let go.

THE AUTOPSY OF A FAILED RELATIONSHIP

I've given my farewells to the past. I've desperately tried to move on. I've walked away. I've learned to release. I let her go. But, I've never conducted an autopsy of the relationship before. I didn't know that it was so time-consuming. I didn't know it was best to perform this procedure at the sunset of my failed relationship rather than two years later. I didn't know that only mature adults were allowed in the operating room because when conductiing an autopsy there is no space for anger, accusation, guilt, or hatred. There's only room for growth, wisdom, strength, and forgiveness. I discovered the hard way, as I usually do, that you can't learn and properly grow from your past if you never take the time to conduct an autopsy of your relationship's death. I have to admit, I tried to suppress my emotions and mask them in duties and accomplishments. But one day or another, you're going to have to dig up that dead corpse and exhume that relationship to discover what really went wrong. You need to do this, not for somebody else, but for yourself. How did you contribute to the death or passing of your relationship? You want to know if you played a role, whether small or large. You want to dig up that corpse because you want to discover the truth. "Was it because of me?" Was it him or her?" "What caused the deterioration and death of our relationship?" You want closure.

For me, I have no ill words to speak of my ex-fiancé. I have nothing bad to say about her. She has a beautiful mind and spirit. Our breakup wasn't about infidelity; there was none. It wasn't about trust; she trusted me and I trusted her. It wasn't about sex; she met me where I asked her to, and she vowed to wait. Simply put, it was about spiritual and emotional immaturity on both of our parts. In the process of surveying the body of our relationship, my responsibility is not to focus or expect for her to see her role, my responsibility is to hone in on my contribution. In being a pastor, I recognize I should have been a champion and better steward over the lot God had given me. I was not. I realize that I have to take onus in the departure of our relationship. I was immature, emotionally and spiritually. As a result, she called it quits. She broke it off with me. She grew tired of my moodiness. She needed somebody who was a little more emotionally consistent and who was a little more financially secure. At the time, I accepted it and moved on, or at least I thought I had.

In my attempt to perform an autopsy, I crafted an email expressing my regrets and apologies for everything I felt I contributed to the relationship and why it didn't work. I sent it. I, then, gave myself a pat on the back because I was being a "big boy." I was taking ownership of my faults. I was being mature. I got no response. Oh wait, yes I did. How could I forget? She sent a short and sweet email letting me know that "Real men don't send messages, they pick up the telephone and call."

"Down goes Frazier!"
"Down goes Frazier!"

My mouth hit the pavement as I read the message she sent. I felt some type of way. That is, I was filled with confusion and raw emotion. I wasn't going to call because that was just too risky for me. I gave myself a passing grade, an "A" for effort. I did what I was supposed to do, right? I tried to apologize and deal with my past, didn't I? I attempted to reach out to her and confess my wrongs. I was trying to be emotionally mature but she obviously wasn't trying to hear it.

When conducting an autopsy of your failed relationship, contrary to popular opinion, you can't focus on the other person. You won't be able to find true healing this way - I've tried it. You can't expect the other person to arrive at the same place of revelation as to how the relationship failed. You have to walk in your accountability, alone. Naturally, we may want to confess our role and we expect the other person to confess their role, but this isn't a true relationship autopsy. If the other person never comes to see how they may have contributed to the demise of your relationship, then you must, alone. You have to be relentless. You also have to operate in grace with yourself. It's okay. Well, maybe it's not okay, but it will be okay, eventually. The best thing you can do is reflect on how you may or may not have played a role. Try not to focus on all the things the other person didn't do and just focus on what you did or could have done better. It's not your responsibility to try and carry the burden of this homicide or suicide investigation, it's your responsibility to find clarity and solace knowing not only the things you did right but the things you may not have done well. You can eventually find peace there. Maybe not now, but soon enough. Accountability can be evasive. It can be elusive. It can be painful and overbearing, but at the same time, it can be peaceful and kind. It bears fruit when you come to accept it for what it is.

Why did your relationship fail? Why did your marriage end? Why did your engagement fall through? Yes, I hear you. I hear all the things about the other person, but what about you? Where do you fit in in this picturesque description of your innocence? I'm not asking that you make yourself guilty, I'm just asking that you take the time for real reflection and try to enter into that bachelor(ette) pad of a blind spot you created for yourself, and be honest with how maybe you failed at something. It's easy to hold onto the past and to collect offenses, convincing yourself that you're the mature one and that it's all on the other person. What would you have done differently? It's okay, just try to discover the beauty of accountability. When it comes to an autopsy of your failed relationships, you have to be able to look at yourself in the mirror and you have to address some real life questions. What did you learn from this relationship that will help you build and foster a new relationship that is healthy moving forward? How do you need to heal from your past relationship? What are some things you need to leave behind with the relationship? Many of us try to take everything with us but there are some things you need to leave there. You have to ask yourself these questions and you have to wait for some answers. Most people have never read their autopsy report. They really don't know what killed their relationship. Many are still wrongfully holding the other person to blame. Let's be clear. It's not about wrongfully taking the blame or finding something that isn't there, but true deliverance and freedom comes when we can reveal the secrets of the heart instead of hiding them from ourselves.

HAVE THE FUNERAL

I'll never forget it. It was one of the first funerals I had to conduct as a pastor. As I entered the funeral home, I picked up the obituary and a small card fell out which read, "Miss me but let me go." Maybe the words rang loud because I found myself betwixt with my past and what was a semblance of my present. I had been trying to move on but I couldn't and I never knew why. I told myself that I was over my past. I quoted all the Scriptures. I sang the hymns. I chanted all the self-help chants. I made the Facebook posts, like everybody does, acting like I was in a better condition than I really was. I did everything I was supposed to do. I guess, deep down I was still hoping. Have you ever "still" hoped before? For this reason, I never had the funeral.

If you want to effectively deal with your past, or if you are going to be able to have the funeral of your failed relationship, you must be willing to become both a physician and a mortician.

As a pastor, I attend funerals pretty often. It's emotionally draining for anybody to have to attend a funeral and to have to endure the process of bereavement, but to have the responsibility of conducting a funeral and to facilitate the burial requires much more of you. It also requires that you tap into supernatural resources. If you want to effectively deal with your past, or if you are going to be able to have the funeral of your failed relationship, you must be willing to become both a physician and a mortician. As a physician, you have to understand that hearts are going to be broken but after assessing the damage, you need to be willing to make the difficult decision of pulling the plug. As a mortician, you're going to have to be willing to do the dirty work of examining the relationship and allowing it to rest in peace. You have to process the relationship, situate your emotions, and dispose of the remains, or the things that may keep you in bondage or try to prohibit you from living freely again. It's time to have the memorial service for your mistakes. There are going to be people who want to focus only on your past and what you may or may not have done right. It's the past. It's time to have the funeral. Stop looking for a heartbeat and searching for a pulse in your relationship. Take off the stethoscope. Make the call. The time of death of your relationship is….. I'm sorry to talk so much about death but when a relationship has failed and is over, it's reminiscent of a funeral. The pain, the memories, the grief, and learning how to walk away, are all things we have to wrestle with when having a funeral. Therefore, there are three things you want to keep in mind when conducting the funeral to your relationship.

1. Do not resuscitate (DNR).

When a relationship has reached its expiration date, you need to stop trying to resuscitate. When you send that text message you know you shouldn't have sent, it's like pulling out the defibrillator or charging your paddles and yelling, CLEAR! Only to get no response. When you see them on social media and it looks like they have moved on so you send a encrypted private message, its like applying mouth-to-mouth and recharging your paddles, yelling, CLEAR! But still, nothing. You've tried everything to bring the relationship back, to revive it. You have to let it go and lay it to rest.

2. Lay the relationship to rest.

The process of laying the relationship to rest is learning how to allow guilt, shame, anger, and unforgiveness to rest in peace, also. You can't expect to have the funeral if you are going to allow guilt to keep you in bondage. Yes, you messed up. Yes, you may have been unfaithful. You may have done a lot of things but when you lay the relationship to rest, you can't keep picking up guilt and wearing the jacket of shame. You can't allow anger and unforgiveness to fester and to keep you from moving on, completely.

> The process of laying the relationship to rest is learning how to allow guilt, shame, anger, and unforgiveness to rest in peace, also.

"47 But even as Jesus said this, a crowd approached, led by Judas, one of the twelve disciples. Judas walked over to Jesus to greet him with a kiss. 48 But Jesus said, "Judas, would you betray the Son of Man with a kiss?" 49 When the other disciples saw what was about to happen, they exclaimed, "Lord, should we fight? We brought the swords!" 50 And one of them struck at the high priest's slave, slashing off his right ear. 51 But Jesus said, "No more of this." And he touched the man's ear and healed him." (Luke 22:47-51, NLT)

After Jesus released Judas from his life, there came a time when Jesus had to come face to face with him again. One of the most awkward things is when you have a random encounter with the person with whom you experienced heartbreak. Jesus encounters Judas and shows us that in the same place of betrayal there can also be healing. Forgiveness is not only essential evidence that you know Jesus, but it is evidence that you have laid that relationship to rest. Like most of us, Peter chooses to fight as a response to Judas' betrayal but Jesus teaches us that to forgive and to seek to heal the person who betrayed you is what truly releases you. Jesus' forgiveness of Judas did not mean they were friends again, nor does you forgiving somebody mean it's time for you to rekindle. Lay the relationship to rest.

3. Bury, don't plant.

There is a difference between burying something and planting something. The difference is the expectation and the motivation. At times we pretend like we are burying a relationship when in fact we are just planting it hoping that it will come back to life and bear us fruit. The danger of planting a dead relationship is that you keep producing more dead relationships. However, if you bury your dead relationship, it will eventually become the fertilizer that will feed the life of a new relationship, when God brings it your way. When you release all the hell, all the heartbreak, and all the bad experiences from your past, it becomes good fertilizer for your future. When you conduct an autopsy and you learn from your past relationships, you avail yourself to become better for your future relationships. Bury, don't plant. You have to keep reminding yourself, bury, don't plant. Bury, don't plant. You don't want to reproduce what you've been trying to let go of; you want God to give you something new and different and fulfilling. Bury. Do not plant.

LEAVE THE GRAVESITE BEHIND

Leaving a relationship is hard. It's accepting the death of something that was once alive and vibrant. *It is the lost friendship that hurts most. Everything else can be easily replaced. That once vital person you shared your life with, a friendship now reduced to mere fading memories of shared experiences, never to be had again.* These are the words written from my ex-fiancé. When my engagement had broken, I made a decision to move on. I didn't stick around or put up much of a fight. I think that's what disappointed her most. I found myself casting my attention in other places and moving directly into another dating scenario. I told myself, there wasn't much to grieve. I felt like I was okay. I didn't seem saddened. I wasn't holding grudges. I was, however, a bit temperamental and emotionally fatigued, but that's it. I left no room for me to feel anything. I was so enamored with "new" and with "different" that I didn't realize what was happening. I wasn't allowing myself to grieve my past relationship. Instead of grieving, I became occupied in another relationship. I put myself in a peculiar position, a position that many who fail to grieve often find themselves; I was dating somebody and still holding onto unresolved issues from my past relationship. Like a water balloon pierced with many holes, I was leaking all over the place. Silently frustrated. Mercurial. Low patience. Unbeknownst to me, my avoidance to grieve was negatively impacting my current relationship. How could I, then, tell the woman I was dating that I moved too fast and that I really needed to take the time to grieve and to process what had happened? Not only would it make me appear weak, vulnerable, and indecisive, and like I was making excuses to flee, but it would have broken her heart, too. I didn't want to accumulate any more broken hearts on my tab. It was my fault because nobody had forced me to move into another relationship.

I teach people that you can indeed lie to yourself. I had become the perfect liar because I didn't wan't to tell myself the truth. I confessed to nobody, but I still had pictures in my storage space. I still had photos on my old computer. I still had photos hidden in my phone. It wasn't that I wanted or needed to see her but I simply wasn't ready to release her from my mind and my spirit. I didn't speak about her nor did I acknowledge my past. I just tucked it away and hoped it would eventually disappear. I had been convicted by God that I was wrong, completely. I needed to learn how to grieve and leave the gravesite behind. It's when we learn to grieve and leave that we are given our lives back. This is the greatest exchange that can take place for you, me, or anybody else. You want your life back. You need your life back. I needed to heal properly but the only way I could do this was by confronting my past, alone. I had to make the difficult decision to break up with the woman I was in a relationship with. God, forgive me.

> It's when we learn to grieve and leave that we are given our lives back.

1. Give yourself permission to grieve.
It's a choice. We find ourselves immediately pursuing relationships because we choose not to give ourselves the space we need to grieve and to reflect. There is always chatter and people and things to distract us and to captivate our attention and to keep us from being alone. We find things to keep us from having to deal with our grief. But it's

okay. It's okay to cry. It's okay to weep. It hurts, I know, but you have to give yourself permission to be sad and to be hurt and to reminisce without apologizing for it. Your relationship was real, so was your love. You can't pretend your relationship didn't exist. However, if you do not allow yourself to grieve, you will never let go. If you never let go, you will always be holding onto some type of hurt or regret or guilt or anger. If you are always holding onto these things, you will never be fully healed. Your healing comes with release. In giving yourself permission to grieve, your grief should also come

In your grief, don't create for yourself a pity party or a room where you collect your offenses and hang onto them like trophies. Your purpose for your grief is to release yourself.

with an expiration date. Although grief is good for you, too much grief is good for nobody. *"I am dying from grief; my years are shortened by my sadness."* (Psalms 31:10) After having his heart broken and his expectations destroyed, David resorted to hiding inside dark caves. In his grief, he confessed that he felt his life wasting away. Your grief should come with an expiration date, whatever that date may be. I can't set the date, neither can you, but you should know that mourning cannot go on forever. In your grief, don't create for yourself a pity party or a room where you collect your offenses and hang onto them like trophies. Your purpose for your grief is to release yourself.

2. Walk away.

Don't let anyone else lie to you. Walking away is the hardest thing you will have to learn to do. Well, for me, it was by far the hardest thing to do. I got no awards for being macho. I didn't receive any certificates for my decision to walk away. Your ability to grieve and leave will determine your ability to heal. After you have accepted, after you have released, after you have conducted your autopsy and done all that you can, after you have made the decision to bury, not plant your relationship, you have to walk away. After you walk away, you don't need to return to the site. If you keep returning to an old relationship, there is a good chance it wasn't ever buried in the first place. Don't get me wrong, there are some relationships that work the second time around, but those relationships are very few. You need to walk away. Many people are unable to enjoy and embrace the beauty of the present because they're still in an exclusive relationship with the pain from their past.

My two-year-old niece taught me another lesson. She taught me how to say, bye-bye. Her first words weren't ma-ma or da-da, but bye-bye. When I would enter into the home, she would greet me with, "bye-bye." I've been trying to teach her other words than these but she seems to be resolved in telling everybody, bye. I know it seems elementary, but sometimes we need to be able to review our past and to simply say, bye-bye.

I've written my own poem.

MISS ME, BUT LET ME GO

When we come to the end of our relationship
And the sun has set for me
I want no rites in a gloom-filled room
Why cry, for the end of this relationship has set you free

Miss me a little but not too long,
And let not anger, regret, or bitterness weigh you low,
Remember the love that we once shared.
Miss it...but you have to let it go

> *For this is a journey that we all must take,*
> *We learn that healthy relationships are not born,*
> *It is something you have to make.*
>
> *As you reflect upon our good times,*
> *Do not be tempted to resuscitate*
> *For this relationship has run its course,*
> *This relationship has reached its expiration date*
>
> *So when you are lonely and sick at heart,*
> *Remember this too shall make you grow.*
> *Bury your sorrow in the Hands of God,*
>
> *Miss me......but now you have to let me go.*

FIVE SIGNS ITS TIME TO MOVE ON

1. Your values don't align.

If all you have in common are "little things," the growth and potential of your relationship is in trouble. There are four "F's" you need to keep in mind: family, faith, finances, and future. Each of these serve as the four legs to the chair on which your relationship should stand. I wish I could tell you that your relationship can thrive off of one leg and survive on only two, but I would not be telling you the truth. If you know that you don't remotely agree in these areas, then it's okay to accept that you probably need to pull the plug. Contrary to popular opinion, sex is not a value and will not keep your relationship happy; it will only delay the inevitable.

2. You're no longer growing.

When was the last time you experienced significant personal growth in your life? Spiritual, Intellectual, Emotional, or Physical growth? You're not at your best when you are with them. They don't challenge you or encourage you to be better. The right person brings out the best out in you. If you're not growing, you're likely not with the right person or your time together has expired. It is crushing and crippling to be connected to somebody who continues to give you what you used to need. You shouldn't be regressing; you should be progressing.

3. You have chosen resignation over rehabilitation.

You have to be careful because disappointment often turns into frustration and frustration into resentment. Resentment turns into anger and your anger grows into rage. Then, rage can transform into indifference. But once you arrive at the threshing floor of indifference, you have arrived at a point of no return. You can be a part of dysfunction for so long that you no longer waste the time to deal with their nonsense. Perhaps you've resolved in your heart that trying to get him to listen to you or to meet you halfway is just not going to happen so in order to keep the relationship afloat you decide it's best to stop fighting. You have heeded the wrong advice if you believe "It doesn't get any better than this." If you were led to believe that dating, relationships, or marriage has to be a struggle and that you have to accept a person's half-hearted attempt to contribute to the relationship, you have believed wrong. If you would rather resign than to try to work on your relationship, then it's time to go. Do you argue so much that you've resorted to mental and emotional resignation? If you have decided to overlook the flaws and the bad habits of your partner because you have concluded that is the best you can do, then it's time to press the eject button.

4. One or both individuals are emotionally absent from the relationship.

For a relationship to soar, both partners must be fully and unequivocally invested. The sucky part of dating is that we often have the tendency to attach ourselves to people who can't reciprocate our affections. As a result, we make excuses for the emotional barriers the other person erects in the relationship to avoid being fully "committed." Many times, we misconstrue personality types with a person's potential to provide emotional intimacy. Because a person has charm, humor, or an engaging personality doesn't necessarily mean they are available emotionally. Just because a person can make you laugh doesn't mean they can meet your needs. It is a terrible feeling to be in the presence of the other person and yet feel as though you are far apart.

5. You no longer talk in detail about your future together.

Be honest with yourself. Does he ever talk in detail about his future with you? You can't allow yourself to ignore this red flag because you're more focused on living in the right now. Any person that desires to be with you and to build a lasting relationship with you will consider the future. Do you talk about children together? Do you talk about how your careers accentuate one another? If he never talks about the future, it's very simple: he doesn't see you in it or he doesn't want to think that far ahead. You want to know, actually, you deserve to know whether he or she is invested in a future with you. If not, you're wasting your time.

CHAPTER 11

BUT WHAT IF MY WOUNDS NEVER HEAL?

LIVING BEYOND YOUR HURT AND BECOMING WHOLE

"How do you pick up the threads of an old life? How do you go on, when in your heart you begin to understand there is no going back. There are some things that time can not mend. Some hurts that go too deep that have taken hold."
Frodo Baggins, Lord of the Rings: Return of the King

While some wounds take longer to heal than others, there are other wounds that may never heal. This is the inconvenient truth. There is some hurt and disappointment, that even after the autopsy, funeral, and grieving process has subsided, still linger. It's those great expectations that fail to come into fruition, which hurt the most and cut the deepest. It's that person who was closest to you that caused you the most devastation. What do you do when those wounds you accumulated just don't seem to heal? How do you take your thoughts captive when you're fast asleep and your mental enterprise chooses to replay the very things you're trying so hard to lay to rest? I felt like Frodo Baggins in the Lord of the Rings. Along his tumultuous journey, he received and carried many scars and wounds, one of which was a stab wound to his shoulder. In a hand-to-hand battle with the Nazgul, a Morgul blade pierced his shoulder and nearly punctured his heart. As a result, he is given medicine that provides him a sense of healing. After some time passes, it is revealed that his recovery is at best, partial, and that he would never be fully healed. A piece of the blade remained in Frodo's shoulder constantly pressing towards his heart and reminding him of his past. Frodo tried his best to live a normal life and to avoid troubling others with his burdens, as he was always thinking of how his afflictions affected those for whom he cared.

My engagement was one of those wounds that just would not heal like the others. This was one relationship that I couldn't easily shake. Like Frodo, I felt a sense of healing, but the recovery was partial and temporary. It felt like I had walked a thousand miles and still had not come close. The autopsy, I performed it. I understood where I went wrong. I embraced my mistakes. I attempted to reach out to her, but she refused it. So why was I still feeling like I was back at square one? Why were my dreams still pervaded with thoughts about her? Why did it feel as if a piece of the Morgul blade was still lodged in me and pressing closer to my heart? Will my wounds ever heal, fully?

THE 9TH STEP

Made direct amends to such a people wherever possible, except when to do so would injure them or others.
-Ninth Step of Al-Anon

I confess without shame that I attend Alcoholics Anonymous meetings. I don't go because I suffer from any known addiction, but because I revere and highly respect the program and the individuals who have the audacity to address their illnesses, diseases, hurts, and flaws, head on. I attend open meetings because I believe the people involved can help us to have the courage to deal with the things that hinder us from becoming the best version of ourselves. How to forgive yourself, how to forgive others, how to start over, how to refrain from premeditating resentment, and how to trust in God, are all the things they teach you and reasons I value this program. It has taught me that recovery is for people who want it, not for people who need it. So I adopted the saying, healing is for people who want it not for people who need it.

There I was, sitting in an open meeting listening to heart-throbbing testimonies and experiencing real brokenness from people sharing their journey to sobriety when I looked over at a board that displayed a worn and fading document called the 12 Commandments of Alcoholics Anonymous. While reviewing each of these commandments, I couldn't help but to notice the Ninth Step that read, "Made direct amends to such a people wherever possible, except when to do so would injure them or others." Then it hit me, I needed to make direct amends with my ex. I read somewhere, "You can't conquer what you won't confront." I needed to confront my past so that I could find peace. Sometimes you might need to find closure, and if possible, you may need to speak with your ex. Maybe you need to feel like you stood up for yourself or maybe you need to clearly articulate what you have always wanted to communicate to them. I was committed to finding and engaging my ex-fiancé. So much so, I did the unthinkable. I did what I coach others never to do. I searched for her on Facebook. I was unsuccessful in my attempt. However, I did find a blog that belonged to her. To my discovery, she had been doing very well with herself. Since our broken engagement she had been traveling the world, seeing places, and doing missionary work. It was kind of like that Julia

> "You can't conquer what you won't confront."

Roberts movie, "Eat, Pray, Love" or something like that. I wanted to see how she was doing and whether she had found new love. I was being nosy. Slowly examining her whereabouts and her temperament, I wanted to know how she was doing spiritually and emotionally. Did she move on? Was she happy? I wanted to know. These are things I wouldn't normally tell anybody but since this is my book, I figured, "Why Not?" I'll tell you the truth.

Over the course of a week, I read her blog like a newspaper to catch me up with the last few years of her life. She was single. She became a nurse practitioner. She appeared happy. But something didn't feel quite right. I remember coming across a recent post entitled, "ONE LESS BELIEVER." It read, *"It only took the actions of one man to destroy the things I used to believe in, or maybe his actions were the final blow to what I'd held onto my whole life…..either way, I don't believe in those things anymore and it makes me very sad."* When I read it, my heart began to hurt because I knew she was talking about me. I saw her hurt and her pain through certain photos. I could see beyond her smiles and her laughter that she was still in some way affected by this breakup, too. The only difference was, I was willing and ready to resolve my issues by any means necessary. According to her blog, she was in town for a few weeks before leaving the country again. It was the 9th Step, and I had an opportunity to make direct amends. I had my doubts. I was trying to talk myself out of it. I was negotiating with myself. I sought counsel from my brother, Jarod. "Isn't it wrong that I want to meet her in person to apologize, if for nothing else, for me?" I asked him. "No," he shouted through the phone in his deep baritone voice. "You need this." So, I began writing her a letter. I was, then, reminded of her last reply to me in an email, "Real men don't send messages, they call." I was trying to find a way of escape so I wouldn't go through with this. The next day, Sunday, I left church and I drove to her house. I waited until the one time no human being should be at home, about eleven o'clock on a Sunday. I felt stalker-ish. I made sure her driveway was clear. I waited until her street had no traffic and then I sped to her house. I quickly jumped out of my vehicle and I dropped the letter in the mailbox and fled like an Olympian sprinter. The letter was very brief. It read,

> *"I know I am the last person you would ever expect to hear from. I heard you were in town for a few weeks and I really wanted to speak with you in person, as you suggested in past. Can you please give me just 3 three minutes of your time? There is a coffee shop right up the street for your convenience. I will be there every morning for an entire week if you should consider to speak with me. I'll be there from 6:30am to 11:30am, every morning."*

For the next 24 hours, the only thing I could think was, healing is for people who want it not for people who need it.

PERMISSION TO BE SELFISH

It is Monday morning and I arrive at the coffee shop just as my letter indicated I would. It's six o'clock in the morning and I have no idea what to expect. So many things flood my mind. I am nervously trying to gather my bearings and create a template of what I want to say. I want to appear sincere, and I don't want her to feel I am trying to waste her time. I feel like I am about to walk into an interview at the White House. I'm sitting here, rehearsing, trying to appear normal and focused when in fact I can't focus at all. I can't keep a single train of thought. Each minute that passes by feels like eternity. It's 6:14am and I am only a few minutes into what feels like a prison sentence. The morning creeps by slowly. I am sitting in this empty coffee shop, alone. I am hoping and waiting for a visitor, and as I quietly expected, she didn't come. I came to get closure and she was a no show. I can't say I blame her. However, when I was preparing to depart, I did notice her car parked across the street. Perhaps she didn't think I was serious? After about ten minutes, she pulled off and went in the opposite direction. I was hopeful. I thought for sure she would show up the next day. The next morning I arrived early and grabbed my coffee and immediately began writing this section of the book. I thought to myself, "This would be a good time to review her blog. Surely, she has written something pertaining to this conundrum." I went to her blog and indeed there was an update from the night before. I hesitantly scrolled down because I knew that whatever she had written would in some way reflect her thoughts about me. It was a bombshell.

The title read, FORGIVE YOU? Her post said, *"Forgive you? Why should I forgive you? To make you feel better about what you have done? Who will make me feel better about what you have done? (Expletive) you."*

When I saw these words, a filthy silence fell on my ears. The coffee shop grew smaller and the oxygen grew thin and it began to slowly exit the room. I couldn't breathe. I became dizzy. I sat there in paralysis for about ten minutes before my heart found its rhythm again. Then slowly, I caught my breath. That was it. Something began feeling extraordinarily different. I was feeling an unusual release. What was this release? I had been seeking to find her so that I could confront

> **NR**
> In order to find true healing you are going to have to risk becoming selfish.

my fears, my shame, my faults, and allow myself to become vulnerable before her. I was trying to heal. I wanted to move forward, but she wasn't ready for that release. I needed to respect that. She didn't want to have a funeral. She was opting to hold onto her anger. We weren't necessarily at the same place.

So, why am I telling you these things? Through this turbulent process, I learned a valuable lesson. In order to find true healing you are going to have to risk becoming selfish. I wrestled with the very thing she was accusing me of. I wondered if I was being selfish. It took me a long time to embrace this fact because I have always been taught that selfishness is a terrible character trait. However, I had to become selfish. I needed this for me. Yes, I wanted her to be better, but even if she wasn't ready or didn't want to, I needed to go through this process for myself. I needed to accept my responsibility, without expecting an apology in return. You won't always get the apology you know you deserve. You can't always get the other person to see how they may have hurt you. That doesn't matter. You must find peace in releasing yourself. She was wrong about one thing. I wasn't trying to meet her so I could "feel" better, I was trying to meet her so I could "become" better.

> **NR**
> Healing isn't happenstance. It doesn't occur by accident. You have to ask for it, seek it, and desire it. You have to think about yourself, unapologetically.

I wasn't concerned about merely feeling good or feeling better about myself. I was trying to discover true healing so that I could help others through similar processes. Healing isn't happenstance. It doesn't occur by accident. You have to ask for it, seek it, and desire it. You have to think about yourself, unapologetically. In the Scriptures, before Jesus heals most people, He asks them their purpose for seeking Him or He asks whether they want to be healed. In either case, each person has to resolve in their minds at that moment that there is nothing else more important than for them to be made whole. They have to elevate their needs over others'.

I showed up to the coffee shop for the remainder of the week anyway. I no longer expected her to come. Sitting there was extremely difficult for me, but I agreed to serve out the rest of my prison sentence there. Walking away from that coffee shop was walking away from three years of holding on. I found my peace and I finally became at peace with my past.

TIME WON'T HEAL YOUR WOUNDS, JESUS WILL

1 After this there was a feast of the Jews, and Jesus went up to Jerusalem. 2 Now there is in Jerusalem by the Sheep Gate a pool, which is called in Hebrew, Bethesda, having five porches. 3 In these lay a great multitude of sick people, blind, lame, paralyzed, waiting for the moving of the water. 4 5 Now a certain man was there who had an infirmity thirty-eight years. 6 When Jesus saw him lying there, and knew that he already had been in that condition a long time, He said to him, "Do you want to be made well?" 7 The sick man answered Him, "Sir, I have no man to put me into the pool when the water is stirred up; but while I am coming, another steps down before me." 8 Jesus said to him, "Rise, take up your bed and walk." 9 And immediately the man was made well, took up his bed, and walked. (John 5:1-9, NKJV)

For ten long years, Bethany harbored ill intentions toward her husband who had physically and emotionally abused her until she was on the precipice of death. He abused her, physically, emotionally, and psychologically. For ten years after their divorce, she avoided him. Although they had children together, she aggressively found ways to keep from ever crossing his path or engaging in any dialogue with him. It was as if he no longer existed. She confessed to me that she had a deep-seated hatred for this man. She went in detail about his abuse toward her and the things he did to destroy her. In those ten years, she never considered the thought of forgiving him. Her wounds were too extensive and they still remained very open. One year into my pastorship, she had grown spiritually by leaps and bounds. One day while sitting down at a lunch meeting, she told me, "I wrote him a letter and I sent it to him. I opened my emotions. I talked about everything I harbored for him, but I also told him that I was choosing to forgive him because I couldn't continue to live with this hatred in my heart. I spent too much of my life hurting because of him." She continued, "But he may never apologize, and I'm okay with that. I had to do it for me. God has been speaking to my heart." I was amazed at her growth. I knew how difficult it was for her to make that decision. She had been carrying these open wounds for ten years and now she was choosing to allow God to close them.

In John 5, there is a man lying helplessly by the Pool of Bethesda when Jesus found him. The Bible says that Jesus saw him and realized that he had been in that condition for a long time. This man had grown accustomed to living his life, sitting by the pool, unhealed. He was accompanied by hundreds of people who were all seeking healing of some sort. This man had been waiting for somebody to come by and escort him into the water so he could be made whole, but after 38 years, his approach had proven ineffective. We don't know if the healing that others received by immersing themselves in the water was temporary, but we do know that it didn't come at the hands of Jesus. This man had been invalid for so long that when Jesus approaches him, He asks, "Do you want to be made well?" One would assume because the man was at the pool, that he wanted to be made better, but Jesus doesn't work off of assumptions. He needed the man to open his mouth and clarify his desires.

Jesus teaches us that not everybody who claims to want to be better and to be healed really, in their heart, desires to do what it takes to be healed. We can be in our condition for so long that we would much rather remain angry and bitter and continue to blame others for our circumstances rather than genuinely seek to be healed. How long have you been living in your condition? Do you really want to be healed? Do you want to get over your past? Do you want to move on or do you want to keep blaming others for the condition you are in?

Jesus saw this man "lying there" hoping that somehow healing would just arrive at his doorstep. When I think about this man who had not been healed and who had been struggling for 38 years, I think about my own life and people I know who have been struggling in their "condition" with a broken heart, unmet expectations, low self-esteem, regret, and unforgiveness for so long. Like this man, we cling to these things for so long and we expect for them to one day vanish. One day turns to six months, and six months turns into three years, and three years turns into twenty, and you are still talking about how that person broke your heart and did you wrong.

Section Three: GETTING PAST YOUR PAST

This man at the Pool of Bethesda:

1. He was looking to the wrong people for help.

When Jesus asked him whether he wanted to be made well, his excuse for remaining in his condition was because there was nobody to help him into the water. He was trying get help from other people who needed just as much help as he did. How long have you been leaving the possibilities of your healing at the feet of another man or woman? How many times are you going to put your expectations in another person to fill your voids so you don't have to deal with your issues? Who are you waiting for in order to be made better?

2. He was depending on the wrong thing to help heal him.

He was expecting the water to provide him the healing that he sought, and it reminds me that so many times in my life, and in life of people close to me, we depend on the wrong things to bring us temporary healing. We adopt coping mechanisms to medicate us so we can mask our pain. Alcohol, prescription drugs, sex, over-eating, retail therapy, attention from others, and so many other things, become the very things we depend on to make us feel better.

3. He had the wrong attitude and perspective.

This man was positioned right by the pool, waiting on something to happen in his life. He was sitting idly, waiting on somebody to walk by and help him. He thought it was enough to arrive at the pool and just wait around. You can move from relationship to relationship, you can arrive to church Sunday after Sunday, you can even lay your problems at the altar, but if you don't possess the right attitude, you will find yourself in the same condition you were in 10 years ago.

Jesus at the Pool of Bethesda teaches us:

1. You cannot heal what you are not willing to unveil.

I imagine this man sitting by this pool for 38 years. He is sitting, lying down, and expecting for someone to come and help him. He thinks someone is supposed to read his mind just because he is sitting at the pool. I want to believe that he didn't engage in much open dialogue with others. He just sits there with his mouth closed and his arms crossed. One thing you must understand, you cannot heal what you are not yet willing to unveil. You can't heal what you are not yet willing to share. Period. If you are going to be made well, if you're going to be healed, if you are going to be made better, you can't keep holding onto what happened to you and expect it to go away. You have to write it down in a journal, you have to confess it, you need to talk about it and not worry about being embarrassed by it.

You cannot heal what you are not willing to talk about and expose.

You can't keep holding onto it, you can't keep quiet. Share your story because your word will be an encouragement to somebody else. *"And they overcame him by the blood of the Lamb and by the word of their testimony."* (Rev. 12:11, KJV) Your testimony will be the reason somebody else will be healed. People will see that you are still here despite what you've been through.

2. You have to give yourself permission to be selfish.

If we were on an airplane sitting together, the flight attendant would instruct us, "In case there is a loss in cabin pressure, yellow oxygen masks will deploy from the ceiling compartment located above you. Please make sure you secure your own mask before assisting others."

You have to give yourself permission to be selfish.

This profound advice transcends flying on a plane; this strategy applies to everything you do in your life especially as it pertains to taking care of yourself. When Jesus asks the man at the pool, "Do you want to be healed?" For one time in his life he got to choose himself. He did not say, "Well, you know the person next to me, they really can use your help. They've been like this for some time." This man chooses himself. You have to say yes to yourself.

3. Time won't heal your wounds, Jesus will.

One thing is for certain, time does not heal all wounds. Whether you are just walking away from a broken relationship or whether you are 10 years removed from an emotionally abusive relationship, you can be just as broken and wounded in year 10 as you were in year 1. Just ask somebody. Time is not what heals, Jesus is the one who heals. This man sat by the pool in the same condition for 38 years and nothing happened. It's not until after Jesus arrives to the scene and asks one question that this man is healed. Jesus did what 38 years couldn't do. This man could not continue to be passive, he had to become more aggressive and direct. If you are looking to be healed, time won't do the job. It's not until you allow yourself to remain in the presence of Jesus that you can allow him speak to your deficits and your deficiencies. Time can't do what Jesus will do for you.

CLOSING THOUGHT

"Some people believe holding on and hanging in there are signs of great strength. However, there are times when it takes much more strength to know when to let go and then do it."
-Ann Landers

"I've learned from personal experience how difficult it is to let offenses go and to turn people loose from the grip of my hands, heart, and mind. In certain points of my life, letting go has been the hardest thing I've had to do. I learned that because I had a fear of letting go, I had been cheating on my future because I was still holding hands and flirting with my past."
-Issac Curry

Pray this prayer:

God, teach me how to release. Teach me how to let go of the anger, hurt, disappointment, and unmet expectations of my past. Teach me how keep my eyes focused on the road ahead instead of my rearview mirror. Give me the awareness so that I can recognize when I am being unfaithful to my future. Give me the intellect and understanding so that I can learn how to let go. Give me the strength so that I can actively apply what I have learned. Give me the discipline so when I walk away, I can focus on what lies ahead. Yes, I have made mistakes but my life won't be defined by my imperfections but by Your grace and mercy. I give my life to You this day.

Amen.

Section 4
CULTIVATING A NEW CULTURE

"You can't start the next chapter of your life if you keep re-reading the last one."
-Anonymous

CHAPTER 12

PRESSING THE RESET BUTTON

LEARNING HOW TO START OVER AGAIN

A woman came home to her husband greeting her with divorce papers after 22 years of what appeared to be a happy marriage. It came to her as a surprise because she couldn't recall any irreconcilable differences or any dissatisfaction in their marriage. He was hard pressed for a speedy divorce and he fought successfully to keep the home they built together. Only one week after she moved out of their house and before she could finish digesting what was happening, her husband's mistress was moving in. To make matters worse, this woman who he had been in a side-relationship with for years was now living among their children. She was beyond devastated. In a written letter to me, all she cared to know was, "How do I start over?"

A man who I am very close with was married for 20 years before his wife suddenly expired, leaving him to pick up the pieces of his life and to have to start over again. A good friend of mine got married only to get a divorce four months later because the person he married literally abandoned him. She left him with no explanation.

How do you handle life when the reset button has been pushed? What do you do? Where do you begin? How do you effectively restore, renew, and rebuild your life after circumstances have dealt you such a fatal blow? Starting over is a spiritual, mental, emotional, and physical process of releasing to God that which we are holding onto. It's the process of opening our hands and releasing our grip on people, desires, outcomes, ideals, feelings, everything. As Melodie Beattie suggests, it's important to acknowledge and accept what we want and what we may want to happen but it is equally important to follow through by letting go. We must let go of our need to live up to the expectations of others. We must let go of our fear of what is perceived to be failure through other people's eyes. In starting over we must refuse to negotiate with reality.

REDEFINING RESET

Reset means something different to everybody. It is a simple word that carries a multiplicity of meanings. Some of us look forward to resetting things in our lives, while others cringe at the sound of the word and the idea of having to do it all over again. When asked, "If you could reset something in your life, what would it be?" One member of my church responded, *"I wish I could reset my thoughts and my attitude. I would take back all the hurtful things I have said to people."* Another member replied, *"I would reset my teenage years because that is where the lies became truth in regards to my self worth and self esteem. Four decades later, I am just now learning who I am in Christ and that I am good enough."* One person said, *"I would reset how I ended my marriage, by cheating."* Sometimes in life, you just need to reset. Reset from the clutter, from your past, from your circumstances, from the chaos. Sometimes you have to walk away, move on, let go, and let bygones be bygones. There are times when you won't be responsible for the reset button being pressed and yet you will be affected by the decision of somebody else and have to start over anyway. Some of us need to press the reset button on a toxic relationship, friendships that are keeping you from growing, bad habits

Section Four: CULTIVATING A NEW CULTURE

and old ways of thinking, and on the fear which has kept you bound from living the life God has called you to live.

I remember the first time I was introduced to the power of a reset button. It came by way of a new gaming system called, Nintendo. This system was very simple in that it came with two basic functions, a power and a reset button. The Nintendo also came with a free game called Duck Hunt. The mission of the duck hunting game was to attempt to shoot as many ducks as possible in order to ascend to the next level. There were times when I would play this game, and even until this day, I feel as if the gaming system were cheating. With my gun literally touching the screen, I would aim and fire directly at the ducks but somehow I would miss

> When it comes to pressing the reset button, I can't confuse my need to start over with my desire to flee whenever something isn't perfect.

a duck. In my competitiveness, I desired a perfect game but when I couldn't get a perfect game, I would resort to immediately pressing the reset button to start the game over again. In the new game, whenever I had missed the mark, in my petty anger, I would resort to pressing the reset button again. For me, this cycle of dysfunction worked.

Whenever the game didn't perform the way I wanted it to, I would simply start over again. Needless to say, I abused the power of the reset button because it became my default when things would not go my way. This same defective behavior was replicated in my relationships with other people. I abused my power of starting over because it became my default and how I learned to respond to conflict. Whenever something would not unfold as I had expected, I would press the reset button. My first option had always been, flight. However, what I am learning is that when it comes to pressing the reset button, I can't confuse my need to start over with my desire to flee whenever something isn't perfect. Sometimes you do have to say, "enough." Sometimes the relationship isn't bringing out the best in you. But sometimes you may need to reset your own attitude, your own behavior, and your own approach to the relationships around you.

FACING YOUR GROUND ZERO

"Face it: we're in a bad way here. Jerusalem is a wreck; its gates are burned up. Come—let's build the wall of Jerusalem and not live with this disgrace any longer. 'I told them how God was supporting me and how the king was backing me up.' They said, 'We're with you. Let's get started.' They rolled up their sleeves, ready for the good work." (Nehemiah 2:17-18, MSG)

My friend Debra has been divorced for several years now. She lives in North Carolina with her two daughters as a music teacher for a private school. She recently shared a secret with me that she had been hiding from everybody for 32 years. Her uncle who is a pastor, molested her when she was a little girl. In her fear and shame, she failed to tell anybody. She grew up never having spoken a word to anybody about this terrible experience. She hadn't spoken to her uncle in 32 years. The depth of this ordeal ran deep, more than I care to share. In her words, this offense contributed to her warped sense of intimacy in her failed marriage. For 32 years she had been suffering from a paralysis from her past. In an attempt to heal from her past, she made the most challenging decision of her life, to speak to her uncle. She was afraid, humiliated, and unsure of the outcome but she felt she could not heal unless she confronted her uncle, in a nonviolent manner.

Section Four: CULTIVATING A NEW CULTURE

She flew halfway across the world to look him in his face and to release her shame, disappointment, and regret. She had erected walls of sexual addiction, codependency, and pride, but after facing her fears she was able to destroy those walls and begin building walls of trust, humility, and a dependence on God.

The book of Nehemiah is an epic story about how to handle life after the reset button has been pressed. Rebuilding the walls of Jerusalem and restoring a people from brokenness, ruin, and despair to a new life with God, the book of Nehemiah serves as an essential tool to help us find our way again. In ancient times, most cities had walls that were strong and tall, which served as a defense system to protect the city. These walls surrounded the city just the way walls surround large castles. After returning from living in captivity, the Jews returned to a city and a temple that was utterly destroyed by their enemies. Jerusalem was in ruins. Nehemiah simulates for us the way of recovery from ruin and despair to restoration and peace.

Take inventory of your life. Maybe your walls of trust and peace have been broken down and barriers of codependency and fear have been erected in its place. As a result, you have very little ability to truly protect yourself. Perhaps you have begun to recycle bad habits that are hard to break and they are prohibiting you from progressing on your journey. Maybe the scars from your past relationship have forced you into hiding and have made it virtually impossible for you to get close to anybody again. Some of you, like Mephibosheth in 2 Samuel 4, were dropped by somebody who was supposed to protect and love you, and, as a result of your incident, you suffered an emotional injury and now you love with a limp. Your gates have been destroyed and nobody recognizes the real you. You're temperamental and calloused and nonchalant. You were robbed of your innocence when you were young and now you set out to steal the innocence of others. Maybe there is a chapter of your autobiography that you don't share with others because that's a part of your life you can't talk about. You desperately try to wipe your memory clean of it. Your gates have been burned and nobody has access to your heart. Most people look at you and they see your glory, but they have no idea of your story. They think you are doing fine but on the inside, there is an increasing and unbearable turmoil brewing. As you examine your walls and your gates, to your discovery, it is in ruins. How do you handle that? This is the dilemma described in Nehemiah.

While in Persia, the first thing Nehemiah does after weeping, fasting, and spending time in the presence of God, he returns to Jerusalem, the destruction site, to assess the damage that had been done. *"After dark I went out through the Valley Gate, past the Jackal's Well, and over to the Dung Gate to inspect the broken walls and burned gates."* (Nehemiah 2:13) Nehemiah was very careful, specific, and intentional in his approach. So should you and I be. We should always be careful and cautious when returning to our destruction site. Nehemiah knew that in order to rebuild something, one must first recognize that reconstruction needs to take place. James Baldwin, an American novelist, said it best, "You cannot fix what you will not face." When pressing the reset button, one of the essential things you must understand after praying, fasting, and spending time in the presence of God, is at some point you must face the ground zero of your past. In order to build on your present and for your future, you must address your ground zero. Nobody wants to return to their destruction site. It's too painful. It brings back too many memories. We want to bury the past so that we don't ever have to deal with those tragic emotions again. We'd rather act like the devastation never took place. The truth is, it did happen.

There is nothing you can do to change the past. You can't run from it, nor can you pretend it never happened. In order to truly make a better life for yourself, you must be able to face what has happened. You don't necessarily have to walk away with answers but you do need to accept what has happened. The most effective way to deal with our hurt is to embrace it and to assess the depth to which the ruin runs. Nehemiah faced his greatest fear - his past. He didn't run from it, but he came to terms with the reality that their lives were a complete mess. Only after surveying the land and doing an appraisal of the damage was Nehemiah able to seek the help of other people in order to rebuild the walls of Jerusalem. He was able to seek help from various workers who accompanied him on the journey. Nehemiah was also successful in receiving special timber from the king's forest before he left Persia, which suggests he was aware of what it would take to effectively rebuild the walls. This means for us that if we are truly concerned about rebuilding parts of our lives, then we need to think seriously about what it will require. We must assess what we will actually need, what steps we should take, and what may be involved in changing our habits, so that we can be able to restore our lives and find peace. For some, it might mean that you need to see a therapist to help you work through some deep-rooted issues. For others, it might mean confronting that person with whom you've held a grudge. Whatever it means, you have to face your ground zero if you want to build again.

> **NR**
> There is nothing you can do to change the past. You can't run from it, nor can you pretend it never happened. In order to truly make a better life for yourself, you must be able to face what has happened.

BACK TO THE BASICS

1. Rebuild your temple.

The most important building in Jerusalem was the temple. The first temple which Solomon built was an impressive building made to house the Ark of the Covenant so the presence of God would have a place to reside and the people would have a place to worship. After the Jews returned to Jerusalem, they rebuilt the temple that was destroyed, but it was much poorer and the people were not as invested in making the temple the center of their affections. Some priests lived in the area while others were neglecting their work as leaders in the community. The people were unfamiliar with God's Law and as a result, they did what was right in their own eyes. Scholars believe that the Jews attempted to rebuild the walls before Nehemiah but they were unsuccessful, mostly because God was not at the foundation of their activities. When you experience heartbreak, disappointment, or devastation of any kind, most times, your worship and your temple suffer the most damage. Our natural reaction is to close God off and to try and handle our misfortunes without God's help. The first step in restoring your life is to restore your relationship, your fellowship, and your time with God. Your mind, your body, and your spiritual walk, are the things you should focus on when trying to reset and start over again. A healthy mind, body, and spiritual life are all byproducts of a healthy temple. Rebuild your temple, first. Begin from the inside out. Your devotional life is paramount, and with this in place, everything else will fall in order.

> **NR**
> The first step in restoring your life is to restore your relationship, your fellowship, and your time with God.

2. Rebuild your gates.

Because the gates were not successfully rebuilt, the Jews were vulnerable to attack. As a result, they compromised with their enemies and allowed their children to marry people from other nations. They had no boundaries nor did they have any standards. It wasn't customary to intermarry with pagans because this encouraged idolatry and prevented the Jews from giving God their undivided attention. If the walls and gates were not rebuilt, the Jews would become weaker and maintaining the temple would have become virtually impossible. If they could just rebuild their walls, they could make the city a safe place again. After rebuilding your temple and repositioning God in the center of your life, you must begin to rebuild your walls. You need boundaries now more than ever before. Boundaries need to be established, and your standards need to be raised. After starting over, you are more likely to attract others who look to take advantage of your brokenness and vulnerability. Without boundaries and standards, we become weaker and it becomes more difficult to maintain our temples.

3. Make your life safe again.

After rebuilding the temple and erecting new gates to protect the city, the Jews needed to learn how to reacclimate themselves to life as Children of God. Before captivity, the people spoke their own language and had their own identity. As a result of the devastation, they lost all these things. While in captivity they had to learn how to speak the language of their oppressors. When the Jews returned to Jerusalem, they needed to learn how to speak their native language again. They needed to be taught how to trust again, and how to make their lives safe. That is, they needed to become the trusting, loving, honest, and fruitful people God made them to be. After starting over, we lose so much in the transition. Sometimes our identity and our ability to communicate is lost in the process of beginning a new life or a new chapter. In the process of starting over, we can become unsafe people who practice unsafe habits. We can become dependent on others. Sometimes, we become cynics who only see the bad in everything and everybody. Sometimes we lose hope and we begin to doubt our self-worth. As a result, we begin to give other people discounts they don't deserve. We cheapen our love and our time, allowing the wrong people to occupy space in our lives. We must learn how to trust and love the right people and this takes time. It's during this season that we must focus on becoming the forgiving, accepting, and graceful people that God desires for us to be. We have to learn that if we carry the bricks from our past relationship into our new relationship, we will build the same house. We must learn how to practice good habits and work on becoming safe people.

CLOSE YOUR GAPS

"But when Sanballat and Tobiah and the Arabs and the Ammonites and the Ashdodites heard that the repairing of the walls of Jerusalem was going forward and the gaps were beginning to be closed, they were very angry, and all plotted together to come and fight against Jerusalem and to cause confusion in it. So we prayed to our God and set a guard as a protection against them day and night."
(Nehemiah 4:7-9 NRSV)

It took hard work, long hours, a lot of praying, and God's help, but Nehemiah and the workers nearly finished rebuilding the walls of Jerusalem, when they experienced their greatest opposition. The governor, Sanballat, who was a believer and follower of God and married into the family of the high priest, became infuriated when they noticed the gaps, or open crevices, in the wall beginning to close. But why would anybody be upset that Jerusalem was being restored and the people were beginning to piece together their lives again? That's a good question. In my life, I've come to experience that there will be people you encounter in your journey, who simply don't want you to rebuild your walls. There will be people who don't want you to close the gaps in your life because those gaps provide areas for them to hide away and live. Some people find comfort in your gaps, and as long as your walls of trust and discipline are down and your gaps are open, there is room for these people to rest.

> As long as you remain with your walls down and gaps open, it keeps people from being accountable for their own lives.

They want to keep things the way they used to be. When people say to you "You've changed," what they are really saying is that you no longer live your life according to their expectations. As long as you remain with your walls down and gaps open, it keeps people from being accountable for their own lives. Some people don't want to see you restored because your being restored forces them to face their own insecurities.

Nehemiah understood that their walls were weak and that they still had gaps in certain areas so they sought to protect these gaps in order to prevent unwanted guests that would breach their security. You have to be careful when you press your reset button. You have to be cognizant of your weaknesses and your proclivities, because these are the areas of your lives that you must secure and protect. If not, you will be inviting the enemy to come in and to threaten the progress you have been making in your life. Whether your gap is an addiction to relationships or sex or wasteful spending or something else entirely, you must recognize, protect, and seek to close the gaps in your lives so that you can be restored. You can miss what God is trying to reveal to you because of a distraction in your life caused by a gap left open and an intruder occupying your attention.

1. Don't be restored

Whenever you are trying to restore your life, discover God's will, or live the life God intends for you, you will always have to deal with chatter from the outside. Chatter from people who say that you don't have what it takes, or that you are not enough. Some people will try to remind you of your past and some will be looking for glimpses of the old you. These people are enemies of your growth. Enemies of growth can be your friends, family, or your coworkers. Like the situation with Nehemiah, your enemies of growth don't want to see you restored. They will seek to discourage, confuse, and keep you from working and focusing on yourself.

2. There is too much rubble

Starting over isn't easy by any stretch of the imagination. Having to explain to children why you're getting a divorce isn't an easy thing to do. Having to return to a home that you both used to share isn't easy either. Especially when you have to come to terms with the fact that the person you've shared so many years of your life with, turns out not to be who you always thought they were. These things require mental, spiritual, emotional, and psychological strength. Yes, we have to be careful and cautious of the people around us who try to keep us from growing. We must also be cognizant of the chatter coming from the inside that tells us, "There is too much rubble." The Jews who were carrying the stones used to build the walls eventually became fatigued. They grew tired of carrying their heavy loads, especially when they looked at how much work they had remaining. The task seemed impossible and they concluded, "There is too much rubble to be moved. We will never be able to build the wall by ourselves" (Nehemiah 4:10). Sometimes we can get overwhelmed and fatigued and feel like we will never be able to pull things together, but this chatter must be ignored. It must be resisted. The enemy wants you to make the mistake of operating and depending on your own resources but you can't fix your situation by yourself. You need the help of God to sustain you.

3. I can't come down

The most classic statement made by any person in the Bible other than Jesus Christ was given by Nehemiah after he committed to rebuilding his life and the city of Jerusalem. Standing on top of the wall, hammering and breaking stones, sweating and toiling in the sun, his enemies devised a plan to distract him from his work. They spread rumors about him and tried to get Nehemiah to react to the gossip spreading throughout the city by stopping and confronting Sanballat to clear his name. When Sanballat's servant approached Nehemiah for a response to all the commotion, Nehemiah replied, "I am engaged in a great work, so I can't come down. Why should the work come to a standstill so I can come down to see you?" (Nehemiah 6:3) This was classic. Nehemiah was so focused on rebuilding his life and becoming better that he didn't have time to address unnecessary rumors or to engage in elementary trifle about things that really didn't matter. You must be so committed to your wellbeing and walking in the will of God that you are unwilling to stop for anybody or anything. Whenever your past comes calling, don't answer it. Send them to voicemail. Why should the good work stop that God is doing through you? You have to resolve that you can no longer lower your standards to come down to another person's level of deficiency.

CHAPTER 13

WHAT'S FOR DINNER?

TAKING SEX OFF THE TABLE AND PURSUING REAL INTIMACY

Growing up in California, I was introduced to sex at the young age of six by an older female friend of the family. This is my second time ever sharing this in my entire life, the first was before the community which I pastor; I shared this with them as I was in the process of writing this book. It was at the age of 14 that I lost my purity, so naturally my view about sex was largely distorted and compromised by my external circumstances. Sex became the operating system from which I learned to function. I was sixteen and a half when I found myself caught in the tangled web of bondage, courtesy of sex. I learned the hard way that how you view sex determines how you use sex. For me, sex was kind of like sitting at a dinner table. Sex became the main course by which I learned to interact and engage in my romantic relationships. Because sex was always on the table, no matter how I spun it and no matter how I looked at it, sex became the driving force that dictated the attitude and the altitude of all my relationships. Because sex was always on the table, I could never really see the other person for who they were. Sex clouded my senses, rerouted my thoughts, and darkened my soul.

Had I known then what I know now, maybe I would have changed course. Had I known that if all you have to offer is sex, you are largely robbing yourself and you don't even know it. If how you communicate has to involve the bedroom, then you have effectively been hoodwinked. It's as if you've put a blindfold around your eyes and walked into a very spacious and visually beautiful room.

    I knew sex was wrong, I was taught sex was wrong, but I wasn't ever taught why sex was wrong. I wasn't taught that God created man and woman to procreate in marriage; therefore, the gift of sex was to answer that particular calling. I wasn't taught that sex was designed to glorify God. When sex was not met with the promise of marriage then the purpose was becoming distorted because it was no longer glorifying God but I was satisfying my flesh. At different points of my life I found myself in bondage having forfeited my freedom because I flirted with lust. Have you ever found yourself in bondage? The more you pray to God, "Lord, I don't want to have sex," the more you find yourself having it. Have you ever found yourself caught in that tangled web and the more you pray the more you seem to fall over that same sin? I learned the hard way that sex blinds you, makes you deaf, and makes you mute. It desensitizes you. It becomes a coping mechanism to hide loneliness, despair, anger, and pride. I know what it's like to have sex rule my life.

IS THE SEX BAD?

I have a best friend whom I've known for 25 years now. In recent years, I have been her sounding board, the person of reason and sound wisdom as it pertains to her romantic relationships. Last year, we were having a conversation about her string of bad relationships and trying to perhaps discover the common denominator in them all. So I took a risk and walked out on a ledge and made a suggestion I knew she would not receive too well. I asked her to consider not having sex and to stop allowing sex to be the feature presentation in her relationships. She responded, "That is not an option." She said, "I love sex too much to deprive myself of it." She paused briefly and then followed, "I need to know what the sex is like because if the sex is bad I cannot remain in the relationship." My reply was woven with sarcasm, "Ohhh. Okay."

You should know something about me, I hate algebra, I have a tasteful disdain for everything arithmetic-related. I have disliked math ever since my seventh grade in Mrs. Lovelace's class. She warned me on the first day of school that if you missed one day, you would always be behind. Unfortunately, I missed one day and I've always been behind. Although I was a complete dweeb in

Sex cannot be the subject and the predicate, noun and the action verb in your relationships.

algebra, it taught me something. It taught me that you can't keep using the same defective formulas to solve the quadratic equations in your life. Although I don't quite remember what "quadratic" means, I think you get the point.

Mrs. Lovelace would give an equation filled with variables, and I was supposed to discover which formula to use in order to solve the equation. My problem was that I decided it was convenient to only remember one or two formulas. I didn't have the time or patience to remember multiple formulas and it showed when I would constantly result in obtaining erroneous answers. I continued to provide the wrong answers because I continued using the same defective formulas to solve my problem.

When I think about my best friend and myself, so many times we have gotten the same wrong answer in relationships because we continued to try and solve our problems of anger, loneliness, bitterness, revenge, and confusion by using the same formula of sex. Sex cannot be the subject and the predicate, noun and the action verb in your relationships. You can't keep using the same defective formulas to answer the quadratic equations in your relationships.

So after a few more relationships, my best friend continued to use her formulas to answer her equations and she continued to receive her same answer: an ended relationship. She finally gave in and messaged me. Her thread of messages read:

- "It's not working out."

- "Why do I keep dating the wrong men?"

- "I can't keep settling."

- "The next time you hear me talking about getting into a relationships, just STOP ME and tell me to think about the last three knucklehead men I decided to get in a relationship with."

- "I always keep picking these certain men."

My simple response,

- "You don't put yourself in a position to thrive."

After each failed relationship, I always ask her,

- "What did you learn?"

During this message conversation, I asked her this question. Her reply:

- "Don't commit too quickly. I should have dated him a little longer to get to know him."

- "I should not have given up my cookies (sex) because that just clouded my judgment and made me feel closer to him than we really were."

- "You can get so caught up in infatuation that you don't take the time to get to learn one another."

Sometime later, in the same relationship she got sick and as a result, she could not have sex for about 4-6 weeks. So for 4-6 weeks, her relationship was void of intimacy via sexual relations. I received a phone call from her and in her disbelief she confessed, "We have nothing in common!" I replied, "Excuse me?" She repeated, "We have nothing in common. When I had to stop having sex, I came to realize we had practically nothing in common. We don't like the same things. We don't value the same things. We stand on two different sides of the spectrum about virtually everything. I can't even get him to go to church. I couldn't see this before and I just can't believe it. I don't understand what I saw in him." She quipped, "Sex had me so wrapped up that I thought we were something more than what we really are. We have absolutely nothing in common and I am so embarrassed."

Like me, for her sex was always on the table in her relationships. However, at the end of that relationship, my best friend finally decided that she had to do things differently. You have to be able to ask yourself, "Besides sex, what do you have in common?" When sex is the direct object of your sentence, you will compromise your values and settle for less than you are worth.

THE MORNING AFTER

"Then suddenly Amnon's love turned into hate, and he hated her more than he desired her."
(2 Samuel 13:15, NLT)

2 Samuel 13 holds within it a very powerful observation and truth for us to consider. There is an obvious and blatant dysfunctional cycle that many, who have not lived under a rock their entire lives, can see. I do not intend to downplay or overlook the heinous act of rape that takes place but before we can arrive there, it is imperative for us to simplify, examine and expose the attitude and the progression of Amnon's emotions and mentality in the text.

1. Man sees woman.

2. Man desires woman so much he worries himself sick.

3. Man thinks he wants to spend his life with woman.

4. Man chases woman.

5. Man gets woman.

6. Man has sex with woman. (There is nothing more to chase after)

7. Man no longer desires woman.

Amnon, David's son, had a half sister named Tamar of whom he didn't have the pleasure of meeting and engaging when they were children because they were raised in different homes. The first time they were introduced to one another, both were adults. When Amnon first met Tamar he fell in love with her. She was so beautiful and soft and meek that he fell head over heels for her. He didn't view her as his family, he saw an ordinary woman that he wanted to make his wife. He desired her so much that he literally made himself sick. Amnon wanted Tamar. He and his friend devised a plan to get Tamar alone in a room with Amnon. The plan was successful. When Tamar got into close proximity to Amnon he forced himself upon her. Although Tamar attempted to reason with Amnon, he overpowered her. She even encouraged Amnon to wait and to at least consider pursuing marriage but Amnon wanted her right then and there. Amnon didn't want to settle for marriage. Instead, he chose to have sex with Tamar; afterwards, the Bible teaches us, *"suddenly Amnon's love turned into hate, and he hated Tamar more than he desired her."*

Sex will delude and convolute your emotions and hijack your rational thinking.

Something must be said about Amnon's attitude before and after he chooses to have sex with Tamar. He thought he wanted Tamar. He thought he fell in love with Tamar. He was head over heels for Tamar; that is, until he had sex with her. After Amnon got what he wanted he no longer desired Tamar. There was nothing else to seek or chase after. There was a great disconnect in how he felt about Tamar before and after he chose to have sex with her.

Many times, we think we want someone until we have sex with them and then that desire quickly fades or becomes overbearing - it never remains the same. After sex, you will find that you are either more attracted or less attracted to the other person than you once were. Sex will delude and convolute your emotions and hijack your rational thinking. Like Amnon, you will find yourself drinking from the chalice of lust and always force-feeding and occupying yourself with sex and never getting a chance to see the other person for who they are. Just because you like a person doesn't mean you need to have sex with them. When sex dominates the relationship, it deceives you into believing you are a match made in heaven when in fact it is more like a date from hell.

After Amnon got what he thought he wanted, the connection was destroyed.

SEX RULES

The word, "rule" can either be a noun or an action verb depending upon how you view the word, sex. You can know that sex outside of marriage is bad for you and it can still find a way to rule your relationships. However, you can also decide to have rules that you choose to govern your life with as it pertains to sex.

When sex rules you:

1. You either commit too quickly or become reluctant to commit.
When sex rules you, there is either an overbearing desire to commit or you become more reluctant to commit. Any way you look at it, the organic flow of the relationship gets disrupted and distorted when you decide to invite sex into your relationship.

2. You begin to mismanage your emotions.
Most women think because you had sex that you both are committed. Some women think that just because you had sex perhaps he feels the same way about you that you feel about him. For some women it is a little more difficult to separate sex and emotions, so sex often

equals some form of relationship and/or promise. For most men, we can somehow suppress and make the divide between sex and our emotions. At least that is what we would like to think we are doing. For most men, sex can easily be viewed as a release. Because of this, emotions begin to get mismanaged when sex becomes a part of the relationship. Sex does not equal commitment. You will find yourself engaging relationships that are meaningless, and you will continue to play the fool.

3. You deprive yourself of clarity.

You deprive yourself of the ability to see clearly, hear clearly, and to operate with clarity in your relationship. Now you want to know, "Where do we stand?" Now you ask, "Where do we go from here?" Now the values you once had are no longer valued. When you decide to be careless and allow sex to rule your relationship, you deprive yourself of clarity. You can't function, think, or make sound decisions.

4. You misappropriate your relationship.

You will find yourself trying to make him or her something they are not. You will find yourself staying somewhere longer and tolerating things you would not normally tolerate because you decided to allow that spiritual, mental, physical, and psychological connection called sex to dominate your relationship. You, like most people, think you have a handle on it but most people don't even realize the real dynamics of sex. Because you casually and carelessly have sex you easily misappropriate your relationships. You put people in positions and allow them to occupy roles that they were never supposed to occupy.

5. You silence the spirit of God.

When you, like I have in my life, get entangled in that web of sex, you start hearing things and then you don't hear things and then you wonder what's going on. You find that you're distancing yourself from fellowship with God. It's practically impossible to immerse yourself in the gift that God has given us, by using it only to satisfy our flesh, without distancing yourself from the voice of God. Now you can't quite hear the voice of God because you're on a different frequency. You have turned down the volume of the spirit of God and turned up the volume of the desires of your flesh. Before you know it, the more you immerse yourself in sex the more you turn down the volume of the voice of God until you just choose to press the mute button.

> **NR**
> Stop thinking that sex is going to make him or her commit because it won't. Sex cannot substitute for substance.

When sex is no longer ruling your life and you learn to have sex rules:

1. You take sex off of the table.

Don't put it down or set it aside. Take sex off of the table - I dare you. When sex is on the table, you think you know the person you are dating. You think you can see who they are, but in reality you only can see a microcosm of the person. You like them because you like the sex. Yep, I said it. You never really interact with the other person because sex is always in the way but when it's not, then you can see them much better. You learn more easily the similarities and differences you have with one another. You discover them and value them beyond the physical and there isn't anything clogging your senses. You can see, hear, and feel. Isn't that amazing? You practice enjoying time and experiences outside of the bedroom. You deserve attention, adoration, and affection that is not connected to the bedroom. But do not make the mistake of believing that just because you take sex off the table your relationship is guaranteed to be successful. It does assure you that you will be clear about the person you are dating. If it doesn't work, it is much easier to break away from a person when you don't have sex keeping you bound together. Stop thinking that sex is going to make him or her commit because it won't. Sex cannot substitute for substance.

2. You learn the value of "defining the relationship"

The natural and more popular sequence of events is: have sex, get emotionally and spiritually attached, try to define the relationship. We introduce sex to the relationship and then afterward we seek to discover "what" and "where" our relationship is going. When you have godly rules to govern your life you learn it is much better and to your advantage to take sex off of the table and also to define your relationship. Are you just casually dating? Are you exclusively committed? Are you just hanging out? You deserve to have clarity and to be able to define your relationship. Sex will not make him commit.

3. You allow yourself the margin to get to know if he/she is God's send to you.

When you learn to take sex off of the table and to define your relationship without it, you allow yourself the ability to determine whether the person you are dating is God's gift to you. You can navigate much better through the relationship when you don't allow sex to disrupt and interfere with the chemistry you are trying to build with the other person. Now that you are here, you have one responsibility, seek God's face to see whether or not the person you are engaging is "the one." Too many times we try to ask this question and seek an answer from God when we are inebriated with lust and overcome with irrationality because of the illusion of closeness and symmetry caused by sex. We then rush marriage because of our misplaced sense of security in this illusion and suffer disappointment and, many times divorce, because we were willfully blinded, bound, and robbing ourselves of due process.

PLAYING WITH DOH

"15 Don't you realize that your bodies are actually parts of Christ? Should a man take his body, which is part of Christ, and join it to a prostitute? Never! 16 And don't you realize that if a man joins himself to a prostitute, he becomes one body with her? For the Scriptures say, "The two are united into one." 17 But the person who is joined to the Lord is one spirit with him."

"19 Don't you realize that your body is the temple of the Holy Spirit, who lives in you and was given to you by God? You do not belong to yourself, 20 for God bought you with a high price. So you must honor God with your body."

(I Corinthians 6:15-16, 19-20, NLT)

When I was young, I was always fascinated with play-doh, a modeling compound used by children for arts and crafts. I loved to use my creative imagination to mold the colorful putty-like substances together to replicate different things. I was a child with many toys and like many children, I was consumed with the fleeting emotion of desiring to play with certain toys only for a season or until another child wanted to play with my toys. When I was introduced to play-doh I remember creating an imaginary monster consisting of various colors. In my professional opinion, my monster was a masterpiece. After some time passed, I decided I wanted to return to my play-doh and create another adventure. To my dismay, I learned I could no longer separate the colors that I previously merged together to make my monster. It was one of the most devastating things to happen to me as a child. I didn't know that once I made the decision to connect different colors, I could no longer undo my decision. I didn't realize at the time that my desire to create something was a temporary decision made with a permanent substance. There were no rules to playing with play-doh, and if there were, I wasn't interested to read them as a child.

Sex, if you didn't know it, is very similar to playing with play-doh. It has permanent implications, whether you and I like it or not. When God created this gift for humankind to celebrate in the likeness of marriage, it wasn't meant to celebrate in the likeness of singleness. The problem is, we've learned to misuse sex and abuse its original purpose given to us by God. When you have sex, beyond the physical closeness, you are merging your spirit with the other person's spirit, thus, making you both become one flesh. This is why the Genesis 2:24 teaches us, "Two shall become one flesh." So, just like my playing with play-doh and after making the decision to merge different colors, I am unable to undo my decision. You and I can't make permanent decisions based on our temporary circumstances.

In the text, Paul uses the example of having sex with a prostitute to try and communicate the gravity of the decision to have sex with somebody other than your wife or husband. It was extreme, I agree, but sometimes you have to take extreme measures to get people to see your point. Paul was trying to get the church at Corinth to understand that when God created marriage, God created things specifically for it. These thing include sharing a home, sharing a bed together, and sexual intercourse. Sex signifies something greater than the here and now. It is more than love, satisfaction, happiness and a release. When you have sex, you are indeed making a spiritual covenant with the other person, similar to the one married couples make.

To better illustrate this principle to my community, I purchased several cans of play-doh representing various colors. In my teaching I sought to disseminate as many pieces of play-doh to as many people who would participate. One person had red, another had blue, and another had a different color. There were about 20 different colors that people were holding in their hands. I began to illustrate that when we meet somebody and choose to have sex with them, we are sharing our spirit with them, "becoming one flesh." I asked each male to find a female and as he engaged her, I asked for each of them to take of piece of their play-doh and give to the other person, thus, representing them giving a piece of themselves to the other person. As they finished, their play-doh was half of their original color and half of the color of the person with whom they pretended to engage in casual sex.

Next, I asked each male participant to find another female participant and after engaging in brief conversation, I wanted them to repeat the act of giving a piece of themselves and share it with the other person. The difference, however, is this time I wanted them to give a little of their original color and a little of the color they gained from their previous partner who they encountered with the other play-doh.

We did this activity a few more times, only to discover that when people looked at their putty-like substances, it no longer looked like it used to look. Everybody's play-doh consisted of different colors, representing the different people they were sharing spirits with when they were having sex with the people they were "dating." Of course, this was an uncomfortable and surreal experience for some as I attempted to shift their paradigm about how sex is much more than a physical experience but it is more spiritual and emotional than anything else.

When I think of Paul's words, I think of my life and my spirit being similar to a single color of play doh. The more I engage one person, I am also unconsciously engaging the other people with whom they have had sex, too. On a different level, this is how sexually transmitted diseases are transferred from one person to another - you are engaging a disease transferred from somebody's previous partner. Try to think of sex like playing with play-doh; you have no control of what you are allowing your spirit to experience when you are casually having sex with somebody else. When I asked my participants about their experience, one said, "If this ball of colorful play-doh represents my spirit then no wonder my romantic life is such a ball of confusion."

Paul reminds the reader that your body is a temple that belongs to God and therefore you and I have been bought with a high price. If our bodies are indeed a house for the Holy Spirit, then it is our responsibility to be hospitable and to provide a clean and clutter-free place for the Spirit to dwell.

SEX AS WORSHIP

I was on a mission trip one day when a friend of mine blew my mind. He said, "You know what Issac? I have a married friend, and you know what he does? He prays before he has sex." There was silence among both of us. The silence was broken by my friend, "That just blows my mind." I was thinking to myself, "That blows my mind too!" My friend is married and I am single but both of us were baffled. I mean, you pray before you have sex? What? Who does that? I thought to myself, "He must be very close to Jesus." I went to sleep that night thinking about this conversation and the question, "How or why would you pray before you have sex?" I woke up the next morning with what I believe to be revelation from God. My conversation with myself went something like this:

Do you say your grace or pray before you eat?

Yes.

Why?

Because God has given me a gift of food and I don't want to take it for granted. I like to express my appreciation to God.

Do you say a prayer before you go to sleep and/or when you wake up in the morning?

Yes.

Why?

Because life is a gift and a blessing.

That was it. The lightbulb went off. So if God gave both husband and wife the gift of sex and as a byproduct, a gift to procreate, then the question should come, "Do you pray before you have sex?" Not a prayer of repentance but a prayer of thanksgiving. Sex is a gift and a blessing; therefore, it should be accompanied with praise, prayer, and thanksgiving. In a sense, sex is a form of worship.

Why can't I fathom the idea of praying before having sex? It is because my view of sex was distorted and blemished. When we can learn to change how we look at sex, we begin to change how we treat sex. When sex becomes a gift for you, given by God to both husband and wife, then sex can become worship for you. When you think of sex as an act of glorifying God then you should be able to pray to God and say, 'thank you' and it should be worship to you. If sex is not worship then perhaps you are not yet correctly seeing it as a gift from God for you.

CLOSING THOUGHT

"It takes ten times as long to put yourself back together as it does to fall apart."
-Finnick Odair, *Mockingjay*

Pray this prayer:

God, I understand that the journey to my destiny is filled with divine detours and unplanned interruptions. Despite my misfortunes, heartaches, and what appear to be setbacks, teach me to find purpose in everything I experience. I can't change the past. Although it may have wounded me, I glory in the fact that even broken crayons still have the ability to color and to create something beautiful. Help me to be reminded that in pressing the reset button, I am not Humpty Dumpty and I can't keep looking to other people to fix me or to put me back together again. I look only to You, God. In You I will find my peace. Help me to walk in my destiny. Help me to endure the roadblocks and the construction sites on the road to achieving the purpose You have for me.

Amen.

Section 5
DATING WITHOUT REGRETS

"We must be our own before we can be another's."
Ralph Waldo Emerson

CHAPTER 14

I KNOW WHO YOU ARE BUT WHO AM I?

REDIRECTING OUR SEARCH FOR THE RIGHT ONE AND BECOMING THE RIGHT ONE

Years ago, I stood on a humble porch of a house in a small town in Kenya Africa, talking with a well respected bishop. I was unveiling past mistakes and mishaps in my dating life. This was before the concept of this book ever came to my mind. Although I've heard this statement many times since this conversation, for me it was a phenomenal paradigm shift. After listening to me go on and on about my desire to be married someday, the bishop looked at me and with his distinguished accent, he said, "Issac, you need to stop putting the onus on other people to be the right person, you need to become the right person for somebody else." I paused. It was like I was in a boxing ring and my opponent discovered I had a glass chin - POW. "Man down!." In a moment's notice, I returned to the conversation. I confess, I heard nothing that came after his initial admonishment. He was talking but I was reflecting. I was glad he had the audacity to speak truth in love to me - sometimes this is what we need most. Be who and what you desire of somebody else instead of expecting somebody to be something you are not yourself.

WHO AM I?

I was in a counseling session when a woman confessed to me that she felt lost without her husband and her marriage because all she did was for him. When they were married, he made the money.

She forfeited her career and her pursuit for graduate education at his request. She stopped dressing up and maintaining her beauty. She had been out of the work-force for so long she confessed that she no longer knew her skill-set or what she could do in order to provide a living for herself. While she was talking, she paused mid-sentence and her eyes began to shimmer with tears. I could tell that what she was about to say was something she didn't really want to share with me. She was embarrassed. She was ashamed. Now that they were getting a divorce, she took a chance and became vulnerable before me. She leaned over and said, "Issac, I don't know who I am." The single most distracting and debilitating thing is to live a life not knowing who you are and what your life's purpose is.

> NR
> The single most distracting and debilitating thing is to live a life not knowing who you are and what your life's purpose is.

Who are you aside from your career? Who are you aside from your nine to five job? Who are you separate from your marriage license and your romance? What is your identity?

Too often we spend most of our time and energy seeking to build our identity with other people and through another person that we fail to invest and secure our identity in Jesus. We forget who we are and we lose ourselves because we become so focused on defining our relationship with somebody else that we are not able to define our relationship and who we are in Jesus Christ. We become so focused on building and branding our relationships, how we want them to look and what we want to achieve in them that we lose ourselves in the process. We waste so much time decorating our relationships for social media that there is very little time to focus on ourselves. "I can't tell you my purpose in life but I can tell you what kind of relationship and marriage I want to have" - that's the unfortunate testimony for many people who would much rather be honest with themselves. We don't know our own life's purpose and yet we run to create purpose with somebody else out because of our insecurities and our loneliness. As a result, we build our relationships around these voids and then later wonder why our relationship is imbalanced. Because we lack identity, we need and depend on other people to give us identity.

When you don't know who you are it becomes so much easier to lose yourself in another person. When you date somebody who is a bit more confident or perhaps has a dominant personality, it becomes easier to submit to the temptation of hiding behind their shadow. Many times it comes in the subtle excuse of "support" or "sacrifice," but the danger in hiding behind another person's shadow is that if you hide long enough, you run a higher risk of becoming who they want you to be rather than who God is calling you to be. If you don't know your life's purpose, then the other person's purpose will inadvertently become yours.

When we depend and rely upon other people to define us, we give them the power to defeat us. God is the one who gives our life definition, not any man or woman. The tension is, however, that we treat God like a drive-thru at a fast food restaurant. We pull up and quickly look through our Bibles. We find only the things that we like. After we find what we like, we place our orders through some quick, haphazard prayer, "Lord, what do you want me to do with my life?" After we make our requests, we pull up to the

> **NR**
> When we depend and rely upon other people to define us, we give them the power to defeat us.

window and then expect God to have our answers ready for us. But we wonder why we are spiritually malnourished and unhealthy and why we don't fully know what God is trying to do in our lives. We won't take the time to focus on God. God is not a drive-thru at a restaurant.

Before there can be a healthy relationship there must be a healthy person or two healthy people. If you desire a healthy relationship you must make sure that you are healthy. When we are healthy, we have an uncanny ability to identify somebody who is unhealthy for us. Just as someone who is focused on exercise and proper nutrition can more easily avoid going through the drive-thru than someone who is accustomed to eating fast food regularly. However, when we are unhealthy, we are blinded and our senses are faint; as a result, everybody and everything around us looks healthy and pleasing to our eyes. A significant faculty in your health is knowing your purpose. This comes when we can learn to embrace our season of singleness.

> **NR**
> If you attach your self-worth to any man or woman, or to the success or failure of any dating relationship or even marriage, you're in trouble. Everything that regret needs to hijack and control your life and relationship is now in place.

If you attach your self-worth to any man or woman, or to the success or failure of any dating relationship or even marriage, you're in trouble. Everything that regret needs to hijack and control your life and relationship is now in place. When you unwittingly hand over your self worth to somebody else, you become a prisoner of pretense, POP. You appear to have value in the other person's eyes when in fact you're not valued as much as you would like to think.

When it comes to developing an identity and understanding who we are, there are three things you must consider:

1. God gives you identity, not any man or woman.

7 Then the LORD God formed the man from the dust of the ground. He breathed the breath of life into the man's nostrils, and the man became a living person. 15 The LORD God placed the man in the Garden of Eden to tend and watch over it. 18 Then the LORD God said, "It is not good for the man to be alone. I will make a helper who is just right for him." 19 So the LORD God formed from the ground all the wild animals and all the birds of the sky. He brought them to the man to see what he would call them, and the man chose a name for each one. 20 He gave names to all the livestock, all the birds of the sky, and all the wild animals. But still there was no helper just right for him. 21 So the LORD God caused the man to fall into a deep sleep. While the man slept, the LORD God took out one of the man's ribs and closed up the opening. 22 Then the LORD God made a woman from the rib, and he brought her to the man. (Genesis 2:7-21, NLT)

God created us to be in communion and relationship with other people but God never created us to be given identity by another person.

God created us to be in communion and relationship with other people but God never created us to be given identity by another person. When God created Adam, he gave him identity and purpose. When God created Eve, He gave her identity and purpose, not Adam. Some of the most miserable and unhappy people I know are those who, married or single, continue to look to another person for their identity. It's easy to live a life in the shadow of other people but you were not created to be someone else's prototype, you were made to be an original. When God created Eve, she was not created to become what Adam wanted her to be but what God designed her to become. You were not made to be what somebody else wants you to be, but who

God created you to be. You have worth and value. That empty feeling; that sense that everybody else has a life and you don't is a lie and a fatal distraction. You have worth and your life is real. Desperately seeking a romantic relationship will not cure or fix a lonely heart because mere loneliness is not the absence of affection or attention, it's the absence of direction, and no man or woman can give you direction or meaning because that comes from God. Let God give your life meaning, not another person or thing.

2. Your identity is intricately connected to your commission.

9 Look! The cry of the people of Israel has reached me, and I have seen how harshly the Egyptians abuse them. 10 Now go, for I am sending you to Pharaoh. You must lead my people Israel out of Egypt." 11 But Moses protested to God, "Who am I to appear before Pharaoh? Who am I to lead the people of Israel out of Egypt?" 12 God answered, "I will be with you. And this is your sign that I AM the one who has sent you: When you have brought the people out of Egypt, you will worship God at this very mountain." 13 But Moses protested, "If I go to the people of Israel and tell them, 'The God of your ancestors has sent me to you,' they will ask me, 'What is his name?' Then what should I tell them?" 14 God replied to Moses, "I AM WHO I AM. Say this to the people of Israel: I AM has sent me to you." 15 God also said to Moses, "Say this to the people of Israel: Yahweh, the God of your ancestors—the God of Abraham, the God of Isaac, and the God of Jacob—has sent me to you. This is my eternal name, my name to remember for all generations. (Exodus 3:9-15, NLT)

Moses struggled with his identity. He wrestled with God's calling upon his life - he wasn't fully sure of his life's purpose. "Who am I?" Moses presented to God. Again, Moses asked, "Who am I,

God?" It was as if God ignored Moses' question and disregarded his inquiries, because God's next statement was about God, not Moses. "God answered, I will be with you, and this is your sign that I AM." It was as if God was saying to Moses, "You can't understand who you are until you first understand who I AM."

In this text, Moses is focused on discovering his identity, but God seems more concerned about Moses' commissioning. Moses didn't realize that his identity was intricately connected to what God was calling him to do with his life - to lead the people of Israel out to Egypt. God's commission is always task-oriented, "Now go, for I am sending you to Pharaoh. You must lead my people Israel out of Egypt." If you're trying to discover who you are, your identity, you must understand that it is connected to God's commission for your life.

Who are you aside from your romantic relationship? What is God commissioning you to do with your life?

For example, God didn't call you to be a nurse, God called you to:

- Save lives.
- Nurture people back to life.
- Help people to transition from this life to the next.
- Introduce people to Jesus even in their season of suffering.

God didn't call me to be a preacher, God called me to:

- Help save lives spiritually and physically.
- Introduce people to His saving grace.
- Nurture people in their brokenness back to emotional and spiritual health.

Your identity is intricately tied to your commission. Who you are is directly connected to what God has called you to do.

3. In order to truly understand who you are, you must understand who God is.

13 When Jesus came to the region of Caesarea Philippi, he asked his disciples, "Who do people say that I am?" 14 "Well," they replied, "some say John the Baptist, some say Elijah, and others say Jeremiah or one of the other prophets." 15 Then he asked them, "But who do you say I am?" 16 Simon Peter answered, "You are the Messiah, the Son of the living God." 17 Jesus replied, "You are blessed, Simon son of John, because my Father in heaven has revealed this to you. You did not learn this from any human being. 18 Now I say to you that you are Peter (which means 'rock'), and upon this rock I will build my church, and all the powers of hell will not conquer it. 19 And I will give you the keys of the Kingdom of Heaven. Whatever you forbid on earth will be forbidden in heaven, and whatever you permit on earth will be permitted in heaven." (Matthew 16:13-18, NLT)

The sequence of events is as follows:

Jesus: who do people say that I am?

Disciples: Some say John the Baptist, some say Elijah, and others say Jeremiah.

Jesus: Who do you say I am?

Peter: You are the Messiah, the Son of the living God

Jesus: You are correct. Flesh and blood did not reveal this to you but my Father who is in heaven.

Jesus: You are Peter, and upon this rock I will build my church....

Jesus asks a question upon which Peter correctly identifies Jesus. Immediately following Peter's correct answer, Jesus turns to Peter and begins to affirm Peter's identity. It is when you learn to confess the absolute lordship of Jesus, God is able to speak into your life and show you who you are. "You are the Messiah," Peter responds. "Now, I say to you that you are Peter (a Rock) and upon this rock I will build my church and all the powers of hell will not conquer it." The greatest transaction takes place when you are able to trust and depend upon God because God, in turn, will trust and depend upon you. Even so, nothing will be able to come against the identity that God gives you, 'not even the powers of hell,' Jesus says.

> The greatest transaction takes place when you are able to trust and depend upon God because God, in turn, will trust and depend upon you. Even so, nothing will be able to come against the identity that God gives you, 'not even the powers of hell,' Jesus says.

Our identity is of the utmost importance. It is so important that Jesus promises to protect it from the powers of hell. Therefore, it is our responsibility to protect our identity from things and other people. If you protect your identity, trust that God will direct your activity.

DESTINY OR BUST

Mark Twain once said, "There are two important days in your life; the day you were born and the day you discover why." However, the tension is that we can live our lives, grow our families, increase our wealth and still never arrive at the "why" because of the interference, chatter, and distractions we have in our lives. "We are born again, of incorruptible seed, the Word of God, which lives and abides in us forever" (I Peter 1:23). Each of us has a destiny: things we are supposed to do; the person we are supposed to become; places we are supposed to travel; and people we are supposed to bless. Just like a seed, God puts destiny inside of us and every day we seek to walk and to please God, destiny grows within us. Although our destiny is always growing within us, Matthew 13:24-25, teaches that, "The kingdom of heaven is like a man who sowed seed in his field. But while everyone was sleeping, his enemy came and sowed weeds among the wheat, and went away. When the wheat sprouted and formed heads, then the weeds also appeared." In this parable, every time the wheat began to grow, weeds began to grow also making it practically impossible to differentiate a wheat from a weed.

Although God has placed destiny inside of you, there is an enemy who seeks to sabotage you by planting weeds that will grow among and around your destiny. Because our adversary, the devil, cannot destroy our destiny he seeks to plant weeds that will grow around our destiny and distract us from obtaining and embracing the things of God. Weeds of fear, discouragement, and disappointment come into our lives. Weeds of divorce, abuse, and addictions begin to grow around our destiny. These weeds make it difficult to see what God is trying to do in our lives. We allow relationships, heartbreak, circumstances, and mistakes to handicap us and to keep us from laying hold of our destiny.

If you know anything about weeds, you know that weeds and grass, or weeds and wheat, often look very similar. For this reason, many times, in our physical lives, we make the mistake of watering our weeds when we were only intending on watering our seeds of destiny. Since our enemy doesn't have the power to destroy what God has for us, our enemy seeks to dangle distractions right next to our destiny so that we will waste our time entertaining the distraction rather than embracing our destiny. Every day somebody is giving up on their destiny because they are nurturing and focusing more on their distractions. You thought you were dating or marrying the person of your destiny when in fact you were wasting time focusing on your distraction. You've been distracted dating the wrong people and waiting on the right person, when you should be trying to focus on becoming the right person for whomever God chooses to place in your life.

The closer you walk with God the more God will be your weed-be-gone. God will shed light on the person that is meant for you but you must first understand that your destiny is not to be found in another person. Your destiny is to be found in the Word of God. Your singular focus should be to become the person that God has designed you to be. You may be on a mission to find the perfect person when you should be allowing God to mature you and grow you and prepare you to be the perfect person for the one God chooses to place in your life.

BECOMING MR. AND MRS. RIGHT

Where do you see yourself? What do you want to do with your life? What is your mission and vision? I was sitting across from my cousin engaging in a much-anticipated conversation about love and relationships. Yes, we are well into our adulthood and I know these questions don't seem relationship oriented, actually they aren't. But there are some things more important than relationships and more important than who you want to marry. You are more important than any of these things.

As he hurried to give a piecemeal response, I immediately interrupted him, "Not a pipe-dream, or something that is not attainable. What is it that you feel you've been called to do with your life?" He looked at me with a perplexed look on his face but his pride wouldn't allow him to say, "I don't know." So for 15 minutes he sat there fumbling with his words and rambling about a bunch of nothing, with all due respect. He then began to talk about his skills and all the things that he knows how to do when I interrupted him, "I don't care about your skill-set. I want to know what it is you want to do with your life?" We went back-and-forth measuring the same scenario two or three times until he looked at me and with all honesty said, "I don't know. Nobody has ever really asked me that question before."

What began as a conversation about his dissatisfaction in his relationship and his displeasure in the performance of his long-time girlfriend, turned into a conversation about him and his satisfaction with himself and the performance in his own life. I didn't scold him or look down upon him; I just wanted to help him change his perspective and paradigm about how he was analyzing his life affairs. He spoke about how she didn't challenge him and how unhappy he was and how he didn't think that she was the person for him. As we sat at my round table in my dining room, I looked over at him and said, "no relationship can ever work if you don't first know what it is you want to do with your life." I also shared with him that as a man, if you are not actively working toward improving your career or accomplishing goals in your life, everybody and everybody around you will suffer. I know this from personal experience.

"Stop looking for the person of your dreams and start becoming someone another person is dreaming about."

This made a difference because the onus of being the perfect person or the person of his dreams had been placed on his girlfriend for some time but he had failed to place the onus on himself. Steven Furtick once said, "Stop looking for the person of your dreams and start becoming someone another person is dreaming about."

He looked at me, very startled, he realized that he needed to think about and focus more on his own life. No relationship is going to work if you don't know what it is that you want to do with your life or who you really are at the core. It's not about money, finances, prestige, or position, it's about fulfillment and it's about fulfilling the goal or the calling that you have on your life.

. My cousin realized he was out of position and operating with the wrong perspective. In the meantime, while you are waiting on the right person, become what you are looking for.

CHAPTER 15

*NO REGRETS: 7 FOOLPROOF METHODS
TO SAFEGUARD YOUR DATING LIFE*

LIVING THE LIFE THAT GOD INTENDS FOR YOU

It's a painful thing, to live in regret. That is, to be held captive by thoughts and actions of old. To wish you had done something more, something different. Living in regret is to be imprisoned by your emotions. "I should have." "I wish I had not." I confess, I have regrets about decisions I've made in my past and for this reason, I can truly speak about the bludgeoning effects your past can have on you. The relationships I've mismanaged. The people I've hurt. Regret is not a fun emotion. So how do we live regret-free? How do we maximize our lives right where we are so that we won't look back and say to ourselves "When I was single, I should have done this or that better."

SEVEN SURE-FIRE METHODS TO HELP YOU TO DATE WITHOUT REGRETS

1. Stop dating in the dark.

I encountered a woman who was married nearly half a century. She was going on and on about how she and her husband were so different from one another. So, I had to know, I asked her, "How did you make it work all these years?" She said something rather remarkable to me. She said, "Roses in the moonlight become dishes in the daylight." This blew my mind. I immediately grabbed my phone as I tried to record these words to my digital note-pad. It rung very deep in my spirit, which made me reflect. What are roses in the dark become dirty dishes you have to clean when the morning comes. For me, this means when it is dark things have a tendency to look good, sound good, smell good, and feel good, but when the daylight comes you look back only to recognize things are not as they seemed and perhaps you have made a bad decision. "What in the world have I gotten myself into?" you ask yourself. So I thought, "Stop dating in the dark." Then I came up with my own saying: "What is bliss and romance in the moonlight becomes dejection and selfishness in the day-light."

When it comes to dating, you want to know what your partner looks like in the daylight, not in the moonlight. The more you direct your relationship toward the daylight of Jesus Christ, the more you are able and capable to see your partner for who he or she really is. You want to date in the daylight. Stop robbing yourself and allowing yourself to be romanced and wined and dined in the moonlight. When it's dark, you have a higher probability of dating their representative and not them. If he or she is the right person, you will have all the time in the world for the moonlight.

If you are trying to avoid dating in the dark, be clear. Let him know where you stand. Provide clarity about your intentions and your expectations. Discuss what "dating" means, or what being "exclusive" means to you. You need to lay the foundation as to the intended direction of this relationship. You also have to remember, the direction and the foundation of the relationship is to be established, not only by what you communicate, but by how you communicate. It ultimately comes down to your actions and interactions with him. Men simply are not attracted to women who try to convince them to be in a more serious relationship. Men are deeply attracted to women who march to the beat of their own drum, who live by their convictions, and who have certain standards when it comes to interacting with men. One of the unspoken standards that is most attractive to a man is a woman who is selective and not necessarily available when it comes to dating men.

When it comes to dating, if you can't be yourself or if you have to date in ambiguity and uncertainty, it is likely that the relationship is not worth your time. Finally, when it comes to dating, take it slow. Another unspoken truth is that the pace of the relationship is typically dependent upon the woman. She dictates how fast or how slow the relationship moves. That is, until she relinquishes this power to the man.

2. Don't date unless you're ready for marriage.

In *Seven Habits of Highly Effective People*, Stephen Covey suggests beginning with the end in mind would help to cultivate a positive outcome for people seeking to reach and to master their goals in life. To begin with the end in mind, for Covey, means to begin with a clear understanding of your future. If you know where you are going and what you are trying to accomplish, it helps you to always take steps in the right direction. In this regard, it is okay to begin with the end in mind when it comes to dating. If you're not willing or ready to be a candidate for marriage, then I would advise you refrain from dating. This will cut down on a lot of drama and unnecessary hurt. Dating is not for fun, contrary to popular belief. If you are looking to have something more established and long-term, you don't deserve to have games played with you. If you are not ready for a committed relationship, something that can evolve into longevity, you should refrain from the dating scene, or from those who are not looking to play games.

I was preaching a sermon series called, "Grey Matter: When things in the Bible are not black and white." I asked my community to submit questions they have long wanted to ask that perhaps nobody was willing to answer. One of the very first questions I had an opportunity to read came from an anonymous member that read, "Is it okay for me to date somebody who is an unbeliever as long as I don't get married to him?" This question arose from the Biblical principle concerning being unequally yoked with unbelievers. My response was very simple. "If your goal in dating is to ultimately be married, why would you waste your time dating somebody you have no intention of spending your life with?" Sometimes we "missionary date." We hang onto people who don't believe as we do, hoping that we are the best chance that our date will become a believer like us. Dating should not be an evangelistic tool and you should not be dating somebody that you don't feel can be a suitor for life partnership.

My initial answer to her question may seem remedial for some, but beyond the reasoning of her question, it remained a relevant inquiry. So many times we date aimlessly and for leisure, wasting other people's valuable time and resources when there are other people who are seeking to date with purpose. If you are not ready for something secure, sober, and steadfast, you should be responsible enough not to put yourself in a role and position with somebody who is expecting this very thing.

3. Cultivate your own garden first.

"Then the Lord placed the man in the garden of Eden to cultivate it and guard it" (Genesis 2:15). Before any matchmaking took place, before there was a suitor for Adam, and before there was a garden for both Adam

and Eve, God asked Adam to cultivate the garden God had given him. We don't know how long Adam took tilling the garden and naming all of the animals in the world, but he did learn to cultivate the garden given him. You have a garden that God has given you and it is your responsibility to be faithful and to make sure the garden you are cultivating is what God desires for you. Learn how to "Choose Me before We." We deny and forfeit our own separateness and individuality because we choose to seek a merged identity with someone else, instead. We recycle these bad habits, which causes us to continue choosing "We" before choosing "Me." In order to produce a healthy relationship, you must first be healthy. Therefore, you need to check yourself before you project yourself. Make sure that you are who God is molding you to be before you try to "be" something with somebody else. We pursue relationships with other people because it is easier than being alone and having to deal with our own issues. As a result, we cling to relationships and eventually we project our issues and problems onto them. We force our partners to become physicians, therapists, firefighters, and caregivers, when in fact, they are only supposed to be our partners. Be who God has called and destined you to be first.

> You have a garden that God has given you and it is your responsibility to be faithful and to make sure the garden you are cultivating is what God desires for you.

4. Believe what you see the first time.

Whenever you go to the movie theater your show usually begins with the "Coming Soon Presentations." During this time, movie producers give you snippets of movies that will be coming to a theater near you. It's a sneak preview of a coming attraction. The point of this marketing tool is to build anticipation for an upcoming film and to give you some idea of what to expect when the movie is released. When you are dating somebody and they show you something about themselves that perhaps alarms you or is a cause for a pause, consider it a "Coming Soon To A Relationship Near You." The late author, Maya Angelou once said, *"When somebody shows you who they are, believe them the first time."* If you are going to date, date responsibly. *"My child, don't lose sight of common sense and discernment. Hang on to them, for they will refresh your soul. they are like jewels on a necklace. They keep you safe on your way, and your feet will not stumble." (Proverbs 3:21-23 NLT)*

Use all of your God-given senses and take nothing for granted. Do not silence your intuition and your discernment; these are gifts from God to use for your protection and for the advancement of God's Kingdom. That queasy, uneasy feeling you're getting is worth praying more about and listening to. You must pay attention to your partner and don't ignore any red flags or try to justify his or her actions. How does he cope with adversity in his life? How does she handle money? You're misleading yourself if you think your circumstances will change the behavior of any individual. Contrary to popular opinion, a ring and a ceremony won't change a person.

If he has anger issues before you get married there is a high probability he will not morph into an extraterrestrial superhero after you get married, so he will likely have anger issues after you jump the broom. If infidelity and trust are obstacles you can't seem to shake before you get married, they will still be there staring you in the face once you sign the marriage certificate. If you noticed flags before you got married, or before you decided to become exclusive, you can believe they will still be there. Do not allow yourself to get caught up in the moment, so much so, that you forfeit your power of awareness. I have sat and counseled countless people who, in retrospect, wished they would have listened to their heart or their eyes. So many people confess how they saw the signs but they wanted to see the good in the other person so they continued forward - only to later live in regret. James T. Draper wrote, "Doubt never means yes and always means no or wait a while: God does not lead through doubt. If you can't get peace, that is an answer." Don't stay in a relationship you'll wish you had abandoned later.

5. Continue living your life.

I mean, let's be honest, once you enter into a relationship, all you want to do is share every single experience together. Although this is true, be careful not to allow the temptation to revolve your world around the other person consume you. Men tend to find it attractive when a woman is not "fully" available with her time, her mind, and especially, her body. Women find it attractive when men are actually accomplishing goals and getting things done in their life. I've said it before and it is worth saying again, the key to dating without regrets and cultivating a successful dating experience is to actually prioritize something else higher than the dating life. This means having values, goals, and plans that don't necessarily involve the other person or require his or her presence in order for it to see fruition. Just because he doesn't like to do something that you love to do, you don't have to give it up. Find a way to keep your passions alive - don't let them die.

Don't be so blinded by romance that you lose yourself and the things you value and cherish most. Create for your relationship healthy boundaries and be intentional to keep some goals and adventures separate from your dating life. If your partner finds this problematic it is probably because they are more controlling or more insecure than they have led you to believe. Don't lose yourself in the other person's life and don't allow the other person to lose themselves in you. If the relationship doesn't work, as least you won't have to hire an investigator or Inspector Gadget to find out where your identity went.

> **NR**
> Don't be so blinded by romance that you lose yourself and the things you value and cherish most.

6. Don't use your past as a crutch to not to commit to your present.

I have a very bad habit of using crutch words when I am speaking, publicly. Crutch words are a collection of words we refer or default to in order to give ourselves more time to think when we are talking. Over time, they become unconscious verbal tics. For example, I have the ugly habit of saying, "You know?" when I really am not even asking a question or "Gotcha" when I really don't even know what's going on. We all use crutch words to help us fill in the gaps in our conversations with other people, especially, when we are unsure what we want to say next or we haven't quite figured out how to position our words.

> **NR**
> Don't use your past as a crutch to not to commit to your present.

Crutch words weaken the point we're trying to make and distort and distract from the purpose of the message. Similar to the crutch words we use every day, we often have a bad habit of using crutch wounds in our dating life. Because we have experienced a failed marriage or relationship, we allow ourselves to use our past as a tool to fear commitment. Over time, our past experiences become excuses and crutches that we use to justify our behavior and the mixed messages we send to others. I know this tool all too well because I used to allow this very thing to hinder and handicap my relationships. We all have had relationships that didn't work out. Some have experienced hurt more than others. Some wounds run deeper and are more flagrant than others, but we all, in some way, have experienced relationships that have failed. Whether it is fear or folly, we can't allow our past to distract us from the purpose that God has for our future. Stop using crutch wounds to fill the gap in your interaction with potentially thriving relationships. Allow your past to be your past and trust God to provide you with a very bright future.

7. You need clear eyes, a filled heart, and opened hands.

If you can't enter your new relationship with a brand new attitude, a renewed sense of peace, and with unbridled expectations, you're not ready to date. As a matter of fact, you're preparing to sabotage what would have or could have been a promising relationship. We often enter into our relationships with crossed eyes, a wounded and half-empty heart, and hands clenched, still holding onto things we should've let go of a long time ago. Because our eyes are crossed, we can't see things and people for who they are, bad or good. When we operate and function with wounded hearts that have little love to give, we enter into the relationship with an expiration date. Sometimes we begin dating relationships with a secret Muay Thai-like grip on our past, but because our hands remain clenched, we disqualify ourselves from being able to receive the blessing right in front of us. Yet, we wonder why we continue to experience disappointment? If you can't open your hands, fully, to receive your new relationship, don't waste your time traveling down that road. If you can't let go of your past mistakes, hurts, and/or the people who may have offended you, you are making a mistake of trying to fill your gap with another person.

In order to date with no regrets, you need to enter each new relationship with clear eyes to see the beauty in the other person, a full heart to love relentlessly, and open hands to receive and embrace a brand new experience.

> **NR**
> If you can't enter your new relationship with a brand new attitude, a renewed sense of peace, and with unbridled expectations, you're not ready to date. As a matter of fact, you're preparing to sabotage what would have or could have been a promising relationship.

FINAL PRAYER

"Neither go back in fear and misgivings of the past, nor in anxiety and forecasting to the future; but lie quiet under His hands having no will but His."
-Elisabeth Elliot (Keeping a Quiet Heart)

Pray this Prayer:

God, how do I learn to keep a quiet heart? How do I direct my emotions so they follow the path You have for them? How do I take my thoughts captive when as soon as I awake in the morning they flee uncontrollably? How to I cancel my faulty service agreements and refrain from scheduling appointments with the department of grievance that keeps me bound in mediocrity and connected to the wrong people? How do I live and date with no regrets?

Teach me how to tame my heart and to control my emotions so they don't control me. Teach me how to protect my covenant so that I won't find myself bound to the wrong person. Release me from the bondage and guilt from my past misgivings. Keep me from worrying about the future, which I don't have any control over. Let me focus on my today and how I can maximize this moment for Your glory. Lead me down the path of purpose into the presence of fulfillment. I am Your instrument and the music I produce with my life belongs to You. Let every tear that I have shed, every mistake that I have made, and every wound that I have received on this journey, become wisdom to guide me in my walk with You.
In Jesus name I pray,
Amen.

ABOUT THE AUTHOR

Issac DeSean Curry

Author. Assistant Pastor. Educator. Missionary. Public Speaker. Preacher

Issac Curry was born in Anaheim, California and later raised in Memphis, Tennessee.

He was reared in a broken home and after experiencing great adversity, at the age of 15, alone and homeless, Issac accepted God's calling into the gospel ministry. He was later ordained.

Issac earned his Bachelor of Science degree from Lane College in Business Marketing and Religion. He earned his Masters of Divinity degree from Princeton Theological Seminary where he focused on New Testament Greek and Practical Theology. He also earned his Masters of Theology degree from Duke University where he focused on Hebrew and Black Church Studies.

Having served congregations in Tennessee, Pennsylvania, Massachusetts, North Carolina, and the Republic of Kenya, Issac recently returned to Memphis, Tennessee by divine direction. Presently, Issac is an assistant pastor of Hope Church, a multiethnic and intergenerational ministry committed to fostering a church for the unchurched in the heart of Memphis. Issac serves as the Singles Pastor and also works closely with the senior pastor to execute special projects.

Issac is an adjunct professor at Memphis Center For Urban Theological Studies where he teaches both Old and New Testament Studies.

God's ministry has taken Issac across the country and around the world, which has fed and influenced his passion to communicate the deep truths of Scripture in a relevant and uncompromising manner to people of all ages, cultures, and ethnicities.

NOTES

David Richo, When The Past Is Present: Healing Emotional Wounds That Sabotage Our Relationships, (Boston, MA: Shambala Publications, 2008), 1, 8, 15.

Buzzfeed, "28 People Who Revealed Their Greatest Insecurities In A Powerful Photography Project," http://www.buzzfeed.com/mackenziekruvant/what-i-be-project-greatest-insecurities#.pjY1QleQ2, October 1, 2014.

T.D. Jakes, 'Transform Your Life With Bishop T.D. Jakes" (from Oprah's LifeClass, May 2014), television interview, Oprah Network.

James MacDonald, "How to Biblically Conquer Your Insecurity" (from the Christian Post Online Magazine, July 2012), http://www.christianpost.com/news/megachurch-pastor-how-to-biblically-conquer-insecurity-77926/, accessed August 21, 2014.

James Baldwin Quotes, http://www.quoteswise.com/james-baldwin-quotes.html.

"The Parent Trap: Setting The Stage For Codependence," http://www.soulscode.com/the-parent-trap-is-the-tk-of-codependence/, accessed September 13, 2014.

"Daily Inspirational Quotes and Sayings," "67 Jim Rohn Quotes," http://www.verybestquotes.com/67-jim-rohn-quotes/, accessed May 21, 2014.

Lewis Smedes, "Forgive and Forget," (New York, NY: Harper Collins Publishers, 1996), x, 5, 27.

Melodie Beattie, The Language of Letting Go: Daily Meditations on Codependency, (Hazelden Foundation, 1990), 351.

Myles Monroe, Waiting and Dating, (Shippensburg, PA: Destiny Image Publishers, 2004), 15.

Peter Michaelson, "The Dire Detriments of Divorce," http://www.whywesuffer.com/the-dire-determinants-of-divorce/, accessed January 3, 2014.

Steven Furtick, "The Greatest Source of My Frustration," a sermon delivered in 2014, at The Elevation Church, Matthews, NC.

Toni Morrison, "Beloved," (New York, NY: Penguin Books), 95.